Building Physics

Marko Pinterić

Building Physics

From physical principles to international standards

Second Edition

 Springer

Marko Pinterić
Maribor, Slovenia

Additional material to this book can be downloaded from
https://www.pinteric.com/books/.

ISBN 978-3-030-67374-1   ISBN 978-3-030-67372-7   (eBook)
https://doi.org/10.1007/978-3-030-67372-7

1st edition: © Springer International Publishing AG 2017
2nd edition: © The Editor(s) (if applicable) and The Author(s), under exclusive license
to Springer Nature Switzerland AG 2021

This Springer imprint is published by the registered company Springer Nature Switzerland AG
The registered company address is: Gewerbestrasse 11, 6330 Cham, Switzerland

Contents

Preface

Civil engineering trends in recent decades have placed increased importance on a multidisciplinary approach in the design, construction and
reconstruction of buildings. Structures are no longer designed only to
provide shelter from natural elements, but they are also meant to establish adequate living conditions and preserve human health. Furthermore,
depletion of fossil fuels combined with climate changes have put building energy efficiency into focus. Although structural integrity remains the
primary concern, knowledge of heat transfer, moisture, sound and light,
which are phenomena traditionally covered by *building physics*, is rapidly
becoming just as important.

However, significant advances in research, standards and legislation are
not matched by advances in building physics education. Many excellent
books cover particular sections of the discipline, often going into great
detail and requiring advanced knowledge of higher mathematics, which
makes them appropriate primarily for physicists. The choice of introductory literature suitable for future civil engineers and architects is much more
scarce. Furthermore, there is a growing gap in terms of concise descrip-
tions of a wide range of phenomena between people working in different
groups and on varied subjects.

When I was entrusted with lectures on building physics for students of
civil engineering and architecture, I decided to tackle this very problem.
I wanted not only to bring all the subjects of interest under one roof but
also to present connections between various topics within building physics, connections between those topics and *physical principles* from which
they derive and connections between theory and application in the form
of *international standards*. Making those connections should make the
topics more instructive and interesting. In addition, following Albert Ein-
stein's aphorism, 'Everything should be made as simple as possible, but no
simpler', my aim is to keep the level of mathematical complexity as low as
possible without distorting physical facts. To help the readers of this book,
I also include introductory parts that deal with physical principles of thermodynamics and wave mechanics while assuming that the reader is familiar with solid and fluid mechanics. Finally, where a dynamical demonstration is essential, I provide a supplementary multimedia content.

My task was greatly simplified by the emergence of well-conceived standards that also cover the symbols and names of physical quantities. Ad-
hering to subject- or group-specific terminology is no longer justified, so
the book thoroughly complies with standardised symbols and names, as
presented in Table A.1.

But just in a few years, significant new developments in international standards have taken place, particularly in noise assessment and daylighting. In addition, a need has arisen to make a more comprehensive treatment of topics such as energy balance, solar gain, ventilation and energy efficient buildings. All of this, together with the aim of addressing perceived shortcomings of the first edition, led to the intention to produce the second edition. Since continued progress in such a vibrant field is undeniable, the important updates will continue to be available online even after the current edition is published.

I am confident that this approach will increase awareness and knowledge about topics related to building physics and help new generations in their professional pursuits. On the other hand, there is always a room for improvement, and I am looking forward to constructive suggestions and criticisms of the book. Please find contact details, book errata, updates as well as supplementary content at http://www.pinteric.com/books/.

Marko Pinterić

Maribor, 13th February 2021.

Introduction

In the process of building design and construction, different disciplines are involved, each with different assignments:

- *Architecture* is primarily concerned with aesthetics and spatial functionality.

- *Civil engineering* is primarily concerned with structural integrity.

- *Building physics* is, on the other hand, primarily concerned with living conditions of occupants and interaction between internal and external environments.

This book covers most important building physics phenomena, which can be categorised into four topics: heat transfer, moisture transfer, sound and light. Despite the extreme diversity of topics, they have many things in common, as shown in Fig. A.

First, we are usually interested in the *transfer of a certain physical entity*, either mass or various forms of energy. The transfer of an entity is best described by relating it to an imaginary frame through which the entity is transferred. The new quantity called *intensity* or *density of flow rate* is an amount of entity that transfers through the frame, divided by the frame area and transfer time. Hence, intensity definition and unit are

$$\text{intensity} = \frac{\text{amount of entity}}{\text{area} \times \text{time}} \quad \left(\frac{X}{m^2\,s}\right).$$

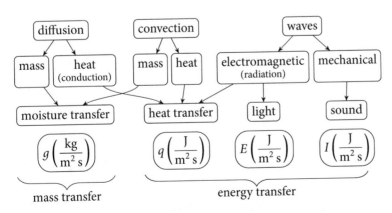

Figure A: Overall picture of physical principles and building physics phenomena covered in this book.

For moisture transfer, heat transfer, electromagnetic radiation (light) and sound, these quantities are called *density of water vapour flow rate g, density of heat flow rate q, irradiance* (*illuminance*) *E* and *time-averaged sound intensity I*, respectively. We will define each of these in the appropriate chapter. However, examining their similarities as well as differences can greatly clarify of their purpose.

Furthermore, processes that facilitate all these transfers are based only on three basic physical mechanisms: diffusion, convection and waves. We will discuss all of these mechanisms as we come to them.

Depending on the problem type, we will study various parts of the buildings [27]:

- A *building element* is a major part of a building such as a wall, floor or roof.

- A *building component* is a building element or a part of it.

We will use those two expressions appropriately throughout the book.

This book also contains an introduction to thermodynamics and wave mechanics for easy reference.

Finally, at the time of publication, the online multimedia contents accompany figures denoted with the symbol shown in the book's margin. Note that more multimedia materials might be added at a later time.

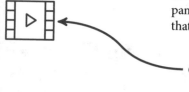

Click here!

1 Basics of thermodynamics

Thermodynamics is the branch of physics that studies bulk properties of systems and the energy transfer between them. By 'systems', we imagine any well-defined regions of the universe under study. This important concept is usually related to 'real' objects, from common solid bodies to gases enclosed within a vessel.

Because the systems usually contain a huge number of particles, they are described statistically. In order to do that, thermodynamics defines additional physical quantities, such as internal energy, amount of substance, pressure and temperature.

The idea of this chapter is not to give a comprehensive and systematic overview of thermodynamics, but a brief introduction to the concepts needed in the rest of the book. However, the chapter ends with the elaboration of energy balance, heat pumps and heat engines.

1.1 Structure of matter

Macroscopic objects always consist of a huge number of particles, that is, atoms and/or molecules. The behaviour of particles depends on the state of the matter (Fig. 1.1):

- In *solid state*, particles are held together very closely by strong electromagnetic forces. The particles can't move around, but they can

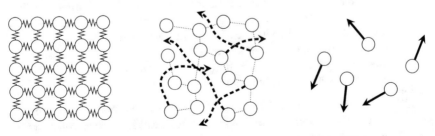

Figure 1.1: Simple models for three states of matter, solid (left), liquid (middle) and gas (right).

© The Author(s), under exclusive license to Springer Nature Switzerland AG 2021
M. Pinterić, *Building Physics*,
https://doi.org/10.1007/978-3-030-67372-7_1

oscillate about their equilibrium positions. Matter in solid state also holds its own shape.

- In *liquid state*, particles are quite close to each other and irregularly connected by weak electromagnetic forces that are easily broken and reestablished. They can move past each other very easily. Matter in liquid state cannot hold its own shape but instead takes the shape of its container. However, because liquid always retains a constant volume, if its volume is smaller than that of the container, liquid forms its own surface.

- In *gaseous state*, particles are separated by a lot of space, and there are no forces between them except for occasional collisions. They move around freely and quickly. Matter in gaseous state cannot hold its own shape but fills the space of its container.

Substances in both liquid and gaseous state flow (continually deform) under an applied pressure difference. Therefore, they are also collectively denoted as *fluids*.

1.2 Heat and temperature

We start the study of thermodynamics by defining a few basic concepts. Some of these definitions will be expanded or elaborated on later, when our knowledge is further extended. We assume that two systems are in *thermal contact* with each other if energy *can be* transferred between them nonmechanically, that is, solely by temperature difference between them. The energy that is transferred from one body to another due to thermal contact is *heat* with unit *joule* Q (J). Note that heat shares the unit with other types of energy. Finally, *thermal equilibrium* is situation, in which two systems are in thermal contact, but no heat is transferred between them.

The first important statement of thermodynamics is the *zeroth law of thermodynamics* or the law of equilibrium: If systems A and B are in thermal equilibrium, and separately systems A and C are in thermal equilibrium, then systems B and C are in thermal equilibrium as well.

This statement can be proved experimentally and is very important because it enables us to define *temperature*. Suppose that we have several different systems. By virtue of the zeroth law of thermodynamics, we can make classes (subsets without common elements) of systems that are in thermal equilibrium with each other, as shown in Fig. 1.2. Then we tag each class with unique temperature T.

Info box

If two bodies are in thermal contact, they can exchange energy in the form of heat. However, if they have the same temperature, the net heat transfer will be zero.

Hence, transfer of heat between two systems in thermal contact depends on their temperature. If their temperatures are different, heat will be transferred between them. If they have the same temperature, no heat will be transferred. In case of radiation, heat is *always* transferred between two systems; however, if both systems have the same temperature, the *net* heat transfer will be zero (Section 2.4.2 on page 59).

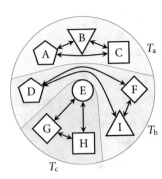

Figure 1.2: Nine systems, with arrows denoting thermal equilibrium. Classes of systems in thermal equilibrium can be formed and 'tagged' with different temperatures.

Using this statement, we can also determine if two systems are in thermal equilibrium without putting them into thermal contact. We can establish that fact using an intermediate, the thermometer, as shown in Fig. 1.3.

As it turns out, temperatures can be quantified. It is possible to develop a method by which a larger number is ascribed to a warmer system, and a smaller number to a colder system. Historically, there have been many attempts to define temperature scales by defining numerical values for two referent conditions. The most prominent scale in everyday use is the Celsius temperature scale, which is denoted by symbol θ (°C). Its unit is *degree Celsius*. The Celsius temperature scale was originally defined by these two referent conditions:

- $\theta = 0\,°C$ is the melting point temperature of water at standard atmospheric pressure (1.013×10^5 Pa).

- $\theta = 100\,°C$ is the boiling point temperature of water at standard atmospheric pressure (1.013×10^5 Pa).

According to this scale, the triple point temperature of water is 0.01 °C.

Later scientific discoveries revealed that in the nature temperature $-273.15\,°C$ is the minimal possible temperature. This temperature, conveniently called *absolute zero*, is the base of the new Kelvin temperature

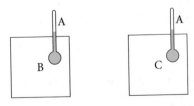

Figure 1.3: The thermometer (object A) is first placed in thermal contact with object B. After thermal equilibrium is reached, we record the first reading. The thermometer is then removed from thermal contact with object B and placed in thermal contact with object C. After thermal equilibrium is reached, we record the second reading. If the two readings are the same, we can conclude that objects B and C are in thermal equilibrium with each other.

scale, denoted by symbol T (K). Its unit is *kelvin*. Note that 1 K represents the same temperature interval as 1 °C. Therefore, the Kelvin temperature scale is defined by these referent conditions:

- $T = 0$ K is the temperature of absolute zero.

- $T = 273.16$ K is the triple point temperature of water.

The temperatures of those two temperature scales can be easily converted using the following expression

$$T = \theta + 273.15. \tag{1.1}$$

Note that the temperature difference/change is the same in both scales

$$T_2 - T_1 = (\theta_2 + 273.15) - (\theta_1 + 273.15) = \theta_2 - \theta_1,$$
$$\Delta T = \Delta \theta. \tag{1.2}$$

> **Info box**
>
> The base temperature scale is Kelvin. However, temperature change/difference is the same in both the Kelvin and the Celsius scale, and both are used to describe it.

Whereas the Kelvin scale is based on a true zero value of temperature, the Celsius scale is based on an arbitrary zero associated with one particular substance at one particular pressure. Hence, in principle, all temperatures should be expressed in kelvins. However, in engineering applications, the more tangible Celsius scale is used whenever possible. Because most processes do not depend on absolute temperature but temperature difference, the Celsius scale appears much more often. To avoid confusion, each scale has a different symbol, so it is easy to resolve which of the temperature scales is used in a particular expression.

1.3 Thermal expansion

One of the best-known temperature-related processes is thermal expansion. When the temperature of a substance increases, its volume usually increases as well. This phenomenon is a consequence of the change in the average distance between the particles in the substance.

In most practical situations, thermal expansion is considerably smaller than the object's dimensions, and the change in any dimension is proportional to the temperature change. This behaviour can be written as

$$\Delta l = \alpha_l \, l_0 \, \Delta T,$$

where l_0 is the initial length along a particular direction, Δl is the relevant elongation and ΔT is the temperature change. The coefficient of proportionality α_l (1/K) is called the *linear expansion coefficient* (Table 1.1).

Because all linear dimensions of an object change with temperature, the object's volume changes as well. The change in volume is also proportional to the temperature change and can be written in analogy to the previous expression as

$$\Delta V = \alpha_V \, V_0 \, \Delta T,$$

Table 1.1: Linear expansion coefficients of most common building-related materials. Presented values are typical or averaged values [12].

Material	$\alpha_l / 10^{-6} \frac{1}{K}$
timber, along the grain	5
timber, across the grain	50
brick	8
gypsum plasterboard †	16
concrete	10
soda lime glass	9
ceramic, porcelain	7
steel	12
stone	8
ice	50

† Values obtained from producers

where V_0 is initial volume, ΔV is volume increase and ΔT is temperature change. The coefficient of proportionality α_V (1/K) is called the *cubic expansion coefficient*.

Taking into account that temperature difference is the same for both temperature scales (1.2), we can rewrite both equations in terms of the Celsius temperature scale as

$$\Delta l = \alpha_l \, l_0 \, \Delta \theta, \tag{1.3}$$

$$\Delta V = \alpha_V \, V_0 \, \Delta \theta. \tag{1.4}$$

Both expansion coefficients must be somehow connected. To derive that relationship, let's consider a cube with an initial edge length of a_0. Note that for isotropic matter in solid-state, the linear expansion coefficient is the same in all directions, so expressions for initial and final volumes are

$$V_0 = a_0^3$$
$$V = a^3 = (a_0 + \Delta a)^3 = \left[a_0 (1 + \alpha_l \Delta \theta) \right]^3$$
$$= a_0^3 \left[1 + 3\alpha_l \Delta \theta + 3(\alpha_l \Delta \theta)^2 + (\alpha_l \Delta \theta)^3 \right] \approx V_0 (1 + 3\alpha_l \Delta \theta).$$

Here we have neglected the smaller higher-order terms because $\alpha_l \Delta \theta \ll 1$ in typical situations. By extracting the volume change and combining it with expression (1.4), we finally get

$$\Delta V = V - V_0 = 3\,\alpha_l \, V_0 \Delta \theta$$
$$\implies \alpha_V = 3\alpha_l. \tag{1.5}$$

The cubic expansion coefficient is especially important for liquids. Because matter in liquid state cannot hold its own shape but instead takes the shape of its container, the linear expansion coefficient does not make sense.

In civil engineering construction, thermal expansion can lead to additional internal stresses. In bridge engineering, an expansion (dilatation) joint is inserted to allow bridge elements to expand without creating those stresses (see Fig. 6.12 on page 206).

1.4 Ideal gas law

As pointed out in Section 1.1, macroscopic systems usually contain a huge number of particles. For example, in 1 kg of water, there are 3.3×10^{25} molecules of water, each of mass 3.0×10^{-26} kg. The relation between the mass of the system m, the number of particles N and the mass of one particle m_1 is

$$m = N\, m_1. \tag{1.6}$$

On the right side of the equation, we have the product of an extremely large and an extremely small number, complicating the use of the expression.

To avoid working with extremely large and extremely small numbers, we can relate the number of particles to a large referent number. We therefore define *mole* (mol) as 6.022×10^{23} particles and *Avogadro constant* $N_A = 6.022 \times 10^{23}$/mol as the number of particles per mol.

Next, we define the *amount of substance* with unit *mole* n (mol) as number of moles

$$n = \frac{N}{N_A}. \tag{1.7}$$

Finally, we define *molar mass* M (kg/mol)

$$M = N_A m_1 \tag{1.8}$$

Info box

The amount of substance and molar mass are handy quantities for describing systems with a huge number of particles.

as the mass of 1 mol of particles. Note that the value of the Avogadro constant is conveniently chosen so that the mass of 1 mol of protons is exactly 1 g.

Putting these two definitions back into equation (1.6), we get

$$m = n\, M. \tag{1.9}$$

Note that amount of substance and molar mass are far more handy numbers than the number of particles and mass of one particle: In 1 kg of water there are 55 mol of water, whereas the molar mass of water is 0.018 kg/mol.

Next we are interested in the macroscopic description of forces that the fluid system exerts on its environment. (We will elaborate on the microscopic background later.) The existence of these forces can be anticipated by observing the inflated balloon, as shown in Fig. 1.4. Because an inflated balloon has much larger dimensions than an empty (relaxed) balloon, forces must exist that keep the balloon membrane in place. The origin of these forces is the air within the balloon. These forces are perpendicular to the membrane, but just as each membrane element has a different orientation, so do the forces. Consequently, it is more appropriate to use scalar quantity to describe these forces. Furthermore, the exerted force on the membrane element is proportional to the element area. We therefore

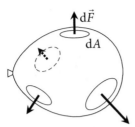

Figure 1.4: Air in an inflated balloon exerts forces on the balloon membrane to maintain its inflated form. Forces are perpendicular to the balloon membrane; hence, they are oriented in different directions, and proportional to the membrane area.

define *pressure* with the unit *pascal* p (Pa) as the ratio of the small area dA to the force exerted perpendicularly to it dF

$$p = \frac{dF}{dA}. \tag{1.10}$$

If perpendicular force F is evenly distributed over the flat area A, we can also use the nondifferential form

$$p = \frac{F}{A}. \tag{1.11}$$

The pascal is equal to N/m^2. Commonly used units are also bar (1 bar = 1×10^5 Pa) and standard atmosphere (1 atm = 1.013×10^5 Pa).

The state of a gas can be fully described by its pressure, volume and temperature. It was experimentally established that for a closed gas with a fixed number of particles, these quantities are related. The simplest relation, valid for most common gases, is

$$p\,V = n\,R\,T, \tag{1.12}$$

where $R = 8.314\,J/(mol\,K)$ is molar gas constant. This is called the *ideal gas law*. In an ideal gas, particles represent only a small fraction of the container; that is, the pressure of the gas is low. Furthermore, particles do not interact between themselves and the container walls, except when they collide elastically.

1.4.1 Dalton's law

Most real gases contain different types of particles. In the most important real gas, dry air, 78.08 % of particles are nitrogen molecules (N_2), 20.95 % oxygen molecules (O_2), 0.93 % argon atoms (Ar) and 0.04 % carbon dioxide molecules (CO_2), whereas the other particles represent less than 0.01 %. Can we use the ideal gas law for such mixtures?

Info box

Air is a gas mixture containing various molecules and atoms.

It turns out that a mixture of ideal gases is an ideal gas, with the total amount of substance n_{tot} and the total pressure p_{tot} as

$$p_{tot}\,V = n_{tot}\,R\,T.$$

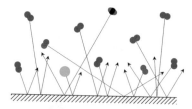

Figure 1.5: Microscopic picture of an ideal gas. The particles within gas occasionally collide with the container walls. Forces caused by collisions with the walls produce gas pressure. Therefore, the forces caused by collisions with particles of a particular component add up to form the total collision force.

In addition, each of the mixture components is itself an ideal gas:

$$p_1\, V = n_1\, R\, T,$$
$$p_2\, V = n_2\, R\, T,$$
$$p_3\, V = n_3\, R\, T,$$
$$\dots .$$

Here we have taken into account that temperature T and volume V are shared by all components of the gas. Pressures $p_1, p_1, p_3 \dots$ are *partial pressures* of individual components of the gas mixture, that is, pressures that the component would exert on the container walls if all other components were absent.

The total number of particles in a gas mixture must be the sum of the particles of the individual components, which can be written in terms of amount of matter as

$$n_{\text{tot}} = n_1 + n_2 + n_3 + \dots .$$

Combining the preceding equations, we get

$$p_{\text{tot}} = p_1 + p_2 + p_3 + \dots . \tag{1.13}$$

The total pressure of the gas mixture is a sum of the partial pressures of the individual components. The statement is known as the *Dalton's law*.

Dalton's law can be understood in terms of a microscopic picture of an ideal gas. Gas consists of many small particles, molecules and atoms, which can freely travel in space. During their movement, they collide with themselves, but they also collide with the container walls (Fig. 1.5). The forces caused by collisions with the walls produce the gas pressure. Therefore, the collision forces with the particles of a particular component add up to form the total collision force.

1.5 First law of thermodynamics

The common misconception relating to thermodynamics is that the system possesses heat. In fact, as we will show, the system can only possess

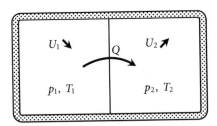

Figure 1.6: The process of internal energy change facilitated entirely by heat, $T_1 > T_2$. Energy is transferred from the warmer system (higher temperature) to the colder system (lower temperature). The internal energy of the warmer system decreases, while the internal energy of the colder system increases. The transferred energy is heat.

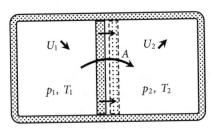

Figure 1.7: The process of the internal energy change facilitated entirely by work, $p_1 > p_2$. Energy is transferred from the expanding system (higher pressure) to the contracting system (lower pressure) regardless of their temperatures. The internal energy of the expanding system decreases, while the internal energy of the contracting system increases. The transferred energy is work.

internal energy and enthalpy, whereas heat and work only measure the transfer of energy between systems.

Individual particles within a macroscopic system can possess mechanical energy. From the description in Section 1.1, we conclude that the particles possess kinetic energy in all states of matter and potential energy in solid and liquid states (electromagnetic force is a conservative force). Because we are interested in a macroscopic description of the system, we define *internal energy U* (J) as a sum of the mechanical energies of all particles. The unit of internal energy is, as for other types of energy, the joule (J).

Internal energy can be changed by two fundamental mechanisms:

1. The first mechanism occurs if two systems are placed in thermal contact, but movement of the barrier between them is disabled, as shown in Fig. 1.6. Energy is then transferred from the system with the higher temperature to the system with the lower temperature, and the transferred energy is heat:

$$\Delta U_2 = -\Delta U_1 = Q.$$

2. The second mechanism occurs if two systems are not in thermal contact, but movement of the barrier between them is enabled, as

> **Info box**
>
> Energy possessed by the system is called internal energy. Heat is the amount of energy that is transferred between two systems.

> **Info box**
>
> Heat transfer tends to equalise temperatures, not internal energies.

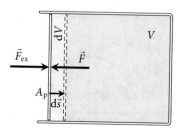

Figure 1.8: Work done by contracting gas. Gas exerts force on a moving piston, which is, by definition, doing work.

shown in Fig. 1.7. Energy is then transferred from the expanding system to the contracting system, and the transferred energy is work:

$$\Delta U_2 = -\Delta U_1 = A.$$

The work that is transferred to or from gas can be calculated. Assume that gas occupies a container, at one end fitted with a piston of area A_p that can be moved without friction, as shown in Fig. 1.8. We slowly (with constant speed) push the piston towards the gas. Two forces, external force F_{ex} and force of gas F, act on the piston and are equal under these conditions. Because gas contracts, the volume change is negative, $dV < 0$. In absolute terms, it is equal to the product of piston displacement and piston area, $-dV = A_p \, ds$. Let's assume that the displacement of the piston ds is so small that the volume and pressure of the gas remain essentially unchanged. Using the definition of work, we can calculate the infinitesimal work done by the external force as

$$dA_{ex} = \vec{F}_{ex} \cdot d\vec{s} = F_{ex} \, ds = \frac{F_{ex}}{A_p} \left(A_p \, ds \right) = -p \, dV > 0.$$

Similarly, we can calculate the infinitesimal work done by gas itself as

$$dA = \vec{F} \cdot d\vec{s} = -F \, ds = -\frac{F}{A_p} \left(A_p \, ds \right) = p \, dV < 0.$$

When gas contracts, $dV < 0$, external work is positive, and the work of gas is negative. However, when gas expands, $dV > 0$, external work is negative, and the work of gas is positive. In the rest of this book we will use the perspective of the gas doing the work, hence

$$A = \int p \, dV. \tag{1.14}$$

In real processes, transferred energy between two systems is a combination of heat and work. We can observe the change in internal energy from the following two perspectives:

1. From the external perspective, the change of internal energy of a system is equal to the sum of the heat supplied from the surroundings to the system and the work done by the surroundings on the system:

$$\Delta U = Q + A_{ex}. \tag{1.15}$$

2. From the system perspective, the change of internal energy of a system is equal to the difference of the heat supplied from the surroundings to the system and the work done by the system on the surroundings:

$$\Delta U = Q - A. \tag{1.16}$$

Both statements are variants of the *first law of thermodynamics*. Note that the differential form of the first law reads the same regardless of the perspective as

$$dU = dQ - p\, dV. \tag{1.17}$$

1.6 Specific heat capacity

When heat is added to the system, the following two fundamental processes can occur:

1. The temperature of the matter increases, so the added heat is called *sensible heat*. On a microscopic level, particles gain kinetic energy.

2. The state of matter changes (phase transition), but the temperature remains the same, so the added heat is called *latent heat*. On a microscopic level, particles gain potential energy.

In this section, we will address the former, whereas the latter will be elaborated on in Section 1.7. It was experimentally determined that added heat is always

- proportional to the mass of matter and
- proportional to *temperature change* $\Delta T = T_2 - T_1$, where T_1 is the initial temperature and T_2 is the final temperature.

We can join these statements into one expression as

$$Q = m\, c\, \Delta T. \tag{1.18}$$

The coefficient of proportionality c $(J/(kg\,K))$ is called the *specific heat capacity*.

Generally, specific heat capacity is temperature dependent, so the preceding expression is more correctly written in differential form as

$$dQ = m\, c\, dT. \tag{1.19}$$

Taking into account that temperature difference is the same for both temperature scales (1.2), we can rewrite equations in terms of the Celsius temperature scale as

$$Q = m\, c\, \Delta\theta, \tag{1.20}$$

$$dQ = m\,c\,d\theta. \tag{1.21}$$

Specific heat capacity is material dependent. Values for a few typical building materials are presented in Table A.3 on page 277.

In engineering, it is more intuitive to assess the relation between the heat and volume of matter. By using the definition of density, we can rewrite (1.20) as

$$Q = V\,(c\rho)\,\Delta\theta, \tag{1.22}$$

where the product of specific heat capacity and density $c\rho$ ($J/(m^3\,K)$) is called *volumetric heat capacity*.

Let's consider the effect of adding heat on a microscopic scale. Note, however, that for the sake of simplicity we have limited ourselves to processes that do not involve phase transitions until the end of this section. First, regardless of the state of the matter, particles move faster, giving rise to kinetic energy. On the other hand, due to thermal expansion (addressed in Section 1.3) distances between particles also increase, giving rise to potential energy in solid and liquid states. Both phenomena increase the internal energy of the system. Additionally, the system also does work due to expansion. For gases, this process can be quantified by the first law of thermodynamics (1.17) as

$$dQ = dU + p\,dV. \tag{1.23}$$

The share of provided heat, which goes either to increase internal energy or to the work of the system, depends on the type of the process. Let's consider the two most important and extreme processes for gases under these conditions (Section 1.9.2):

1. An *isochoric process* has a constant gas *volume*. The coefficient of proportionality for such a process is called *specific heat capacity at constant volume c_V*, as in

$$dQ = m\,c_V\,dT. \tag{1.24}$$

   Because volume change is $dV = 0$, gas does no work, and all provided heat goes to increase internal energy. From (1.23), we get

$$dU = m\,c_V\,dT. \tag{1.25}$$

   The importance of this equation is that we have for the first time established the direct relation between internal energy and temperature. The rise in internal energy is usually (but not necessarily) connected to the rise in temperature.

2. An *isobaric process* has a constant gas *pressure*. The coefficient of proportionality for such a process is called *specific heat capacity at constant pressure c_p*, as in

$$dQ = m\,c_p\,dT. \tag{1.26}$$

In that case, provided heat goes to increase internal energy as well as to the work done by the system. From (1.23) and (1.25), we get

$$dQ = m\, c_V\, dT + p\, \frac{dV}{dT}\, dT = m\, c_p\, dT$$

$$\implies c_p = c_V + \frac{p}{m}\frac{dV}{dT}.$$

For ideal gases, we can use (1.12) and (1.9) to obtain

$$c_p = c_V + \frac{p}{m}\frac{nR}{p} = c_V + \frac{R}{M}. \tag{1.27}$$

Specific heat capacity at constant pressure is larger than specific heat capacity at constant volume. The heat that has to be provided to increase the temperature at constant pressure is larger because it is spent to increase internal energy of the system *and* to do work to expand the system.

The latter process is very common, so we define *enthalpy H* (J), which is, like internal energy, the attribute of the system. Similar to (1.25), we write

$$dH = m\, c_p\, dT. \tag{1.28}$$

Enthalpy accounts for both the system's internal energy and the energy needed to make space for that system.

> **Info box**
>
> Enthalpy accounts for both the system's internal energy and the energy needed to make space for that system.

Example 1.1: Difference between specific heat capacities.

Calculate the difference between specific heat capacities for concrete at atmospheric pressure $p_{atm} = 1.013 \times 10^5$ Pa. Take the density of the concrete to be $\rho = 2.2 \times 10^3 \, \mathrm{kg/m^3}$, linear expansion coefficient to be $\alpha_l = 1.0 \times 10^{-5}\, \mathrm{K^{-1}}$ and specific heat capacity at constant pressure to be $c_p = 1000\, \mathrm{J/(kg\,K)}$.

Heat added at constant pressure (1.26) is spent to increase the internal energy and do work to expand the material (1.23):

$$Q = m\, c_p\, \Delta T = m\, c_V\, \Delta T + p\, \Delta V.$$

We can calculate the volumetric expansion by using expressions (1.4) and (1.5):

$$\Delta V = \alpha_V\, V_0\, \Delta T = \alpha_V\, \frac{m}{\rho}\, \Delta T = 3\frac{\alpha_l\, m}{\rho}\, \Delta T.$$

Putting both expressions together, we get the difference between specific heat capacities:

$$m\, c_p\, \Delta T = m\, c_V\, \Delta T + 3\, m\, \frac{p\, \alpha_l}{\rho}\, \Delta T$$

$$c_p - c_V = 3\frac{p\, \alpha_l}{\rho} = 1.38 \times 10^{-3}\, \frac{\mathrm{J}}{\mathrm{kg\,K}}.$$

Note that the difference between specific heat capacities is six orders of magnitude smaller than the specific heat at constant pressure:

$$\frac{c_p - c_V}{c_p} = 1.38 \times 10^{-6}.$$

The vast majority of materials in solid and liquid states have small linear expansion coefficients and large densities. As a consequence, the provided heat that is spent to expand materials is much smaller than the heat that is spent to increase the internal energy, so the former can be simply neglected. Therefore, for solids and liquids the difference between specific heat capacities is negligible as

$$c_V \approx c_p = c.$$

Therefore, for materials in solid and liquid states (Table A.3 on page 277), only one specific heat capacity is provided.

1.7 Phase transitions

In exceptional situations, (1.20) is not valid; despite heat is being added to the system, the temperature remains constant. The most important exceptions are phase transitions, that is, changes between solid, liquid and gaseous states.

To understand the phase transition process better, let's observe a procedure in which we slowly heat up water from $-30\,°C$ to $130\,°C$ at constant, standard atmospheric pressure. The process is graphically presented in Fig. 1.9.

- In case A, $\theta_A < 0\,°C$, all water is in the solid state (ice). By adding heat, the temperature of the water increases.

- In case B only, $\theta_B = 0\,°C$, water starts melting. By adding heat, the water transforms from the solid to liquid state, *but the temperature does not change until the transform is complete.*

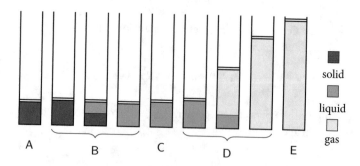

Figure 1.9: Process of heating up water from $-30\,°C$ to $130\,°C$ at constant, standard atmospheric pressure. The presented cases correspond to five temperatures, $\theta_A < 0\,°C$, $\theta_B = 0\,°C$, $0\,°C < \theta_C < 100\,°C$, $\theta_D = 100\,°C$ and $\theta_E > 100\,°C$.

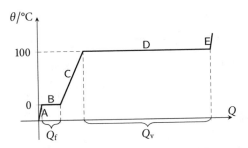

Figure 1.10: Heat diagram for warming water. The horizontal lines correspond to the phase transitions for which temperature is constant. In the rest of the graph, temperature change is proportional to heat.

- In case C, $0\,°C < \theta_C < 100\,°C$, all water is in the liquid state. By adding heat, the temperature of the water increases.

- In case D only, $\theta_D = 100\,°C$, the water starts boiling. By adding heat, the water transforms from the liquid to gaseous state, *but the temperature does not change until the transform is complete.*

- In case E, $\theta_E > 100\,°C$, all water is in the gaseous state (vapour). By adding heat, the temperature of the water increases.

Note that water in the solid and liquid states coexists in thermal equilibrium only in case B, at temperature $\theta = 0\,°C$. On the other hand, water in the liquid and gaseous states coexists in thermal equilibrium only in case D, at temperature $\theta = 100\,°C$. We can therefore define the *melting temperature* as the temperature at which solid and liquid states are in equilibrium. Similarly, the *boiling temperature* is the temperature at which liquid and gaseous states are in equilibrium.

The temperature versus heat plot for the process of heating up water is drawn in Fig. 1.10. The inclined lines in the diagram (A, C and E) are described by (1.20). Note that specific heat capacities for solid, liquid and gaseous water differ, creating the different slopes of the inclined lines.

The length of horizontal lines (B and D) represent the heat required to melt all solid water, which is called heat of fusion Q_f, and the heat required to vaporise all liquid water, which is called heat of vaporisation Q_v. These heats are in fact proportional to the mass of the matter, so we can write

$$Q_f = m\, q_f, \tag{1.29}$$
$$Q_v = m\, q_v, \tag{1.30}$$

where $q_f\,(J/kg)$ is the *specific heat of fusion*, and $q_v\,(J/kg)$ is the *specific heat of vaporisation*. Specific heats of fusion and vaporisation are the amounts of heat required to melt or vaporise the matter per the mass of matter, respectively.

The expressions (1.29) and (1.30) are written from the perspective of heat transfer to the system. Note that by providing the energy, bonds between particles are broken, and potential part of the internal energy is increased. On the other hand, phase transitions are usually accompanied by matter

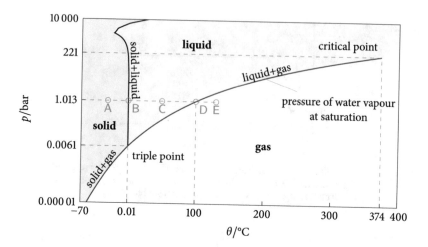

Figure 1.11: Phase diagram of water. The red line is water vapour pressure at saturation. Points A, B, C, D and E correspond to the cases in Fig. 1.9.

expansion, which means that work is done by the system. From the perspective of the system itself, the provided heat is equal to the change of enthalpy as

$$\Delta H_f = m\, h_f, \tag{1.31}$$

$$\Delta H_v = m\, h_v, \tag{1.32}$$

where h_f (J/kg) is the *specific enthalpy of fusion*, and h_v (J/kg) is the *specific enthalpy of vaporisation*. Obviously, the specific heats and the specific enthalpies are the same quantity, $q_f = h_f$ and $q_v = h_v$.

The difference between more common equations (1.29) and (1.30) on the one hand, and more peculiar equations (1.31) and (1.32) on the other hand, is that the former pair describe the phase transition processes from the perspective of the energy transfer, whereas the latter pair describe phase transition processes from the perspective of the system energy change. Because our primary interest is system properties, we will use the latter ones in the rest of this book.

The state of matter depends on the temperature of matter. However, water at room temperature can also be in the gaseous state at a very low pressure and in the solid state at a very high pressure. The state of matter dependence on pressure is taken into account in a pressure versus temperature plot, called a *phase diagram*.

A phase diagram for water is shown in Fig. 1.11. The shaded areas represent individual states of matter—solid, liquid and gas—whereas the boundaries (lines between the areas) represent two states of matter in equilibrium. Especially interesting is the point at temperature 0.01 °C and pressure 0.061 bar at the junction of all three lines. This point, in which the solid, liquid and gaseous states are in equilibrium, is known as the *triple point*.

The line that represents the boundary between liquid and gas ends at temperature 374 °C and pressure 221 bar. This point is known as the *critical point*. For pressures and temperatures larger than the pressure and temperature of the critical point, we can no longer distinguish between the liquid and gaseous states, and the transform between them is seamless.

The six possible transforms are

1. *melting*, from solid to liquid;

2. *freezing*, from liquid to solid;

3. *vaporisation*, from liquid to gas;

4. *condensation*, from gas to liquid;

5. *sublimation*, from solid to gas; and

6. *deposition* from gas to solid.

The most important boundary is the one between the gaseous state and the remaining two states of matter. The function that describes this boundary is called *water vapour pressure at saturation* and is used to describe the air humidity and prediction of condensation/deposition.

Water vapour pressure at the saturation line shows that water at a higher pressure boils at a higher temperature. This effect is used in pressure cookers, which increase the pressure and, consequently, the temperature of the water to make cooking faster.

> **Info box**
>
> Water at a higher pressure boils at a higher temperature.

Note that water vapour is only a small component of humid air. From Dalton's law we conclude that the water vapour pressure must be much lower than the atmospheric pressure and that water can vaporise at temperatures much lower than 100 °C. This process is called evaporation and will is discussed in Section 4.2 on page 120.

1.8 Building energy balance

In Section 1.5 we have learned that energy is spontaneously transferred from the system with the higher temperature to the system with the lower temperature, and that the transferred energy is called heat. The undesirable consequence of this process is that a building loses energy during the winter (cold) period and gains energy during the summer (warm) period, as shown in Fig. 1.12 (a). In order to keep the temperature inside the building constant, we need to replace the energy lost during the winter period and dispose of the energy gained during the summer period, that is, to establish the energy balance.

However, it is impossible for heat to flow spontaneously from the system with the lower temperature to the system with the higher temperature. This obvious fact is the consequence of the second law of thermodynamics, which we will also study in Section 1.9.1. Nevertheless, there are several viable strategies for achieving an energy balance, as discussed below.

One possible strategy to replace the lost energy is the use of *furnace*, as shown in Fig. 1.12 (b). Chemical energy in the form of fossil fuels (heating

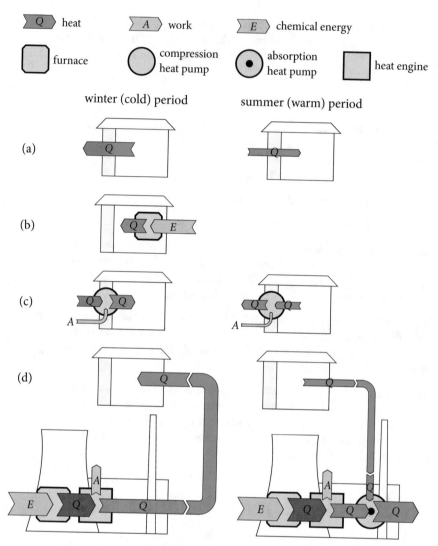

Figure 1.12: Building energy balance. The energy lost during the winter period can be replaced by a furnace, a heat pump or in conjunction to a CHP system. The energy gained during the summer period can be disposed of by a heat pump or in conjunction to a CCHP system. Red and blue tones indicate the temperature at which heat is transferred, with the red and blue tones corresponding to the temperature above and below room temperature, respectively.

oil, gas, coal) or biomass (wood) is first fed into the furnace, where it is converted into internal energy by combustion. This energy is then supplied to the building in the form of heat. The energy conversion is almost perfect, so that the efficiency of the furnace is almost 100 %.

Another possible strategy for establishing the energy balance is the use of the *heat pump*. A heat pump transfers heat from the lower temperature to the higher temperature, which seems to contradict the second law of thermodynamics. However, heat can flow from a lower to a higher temperature without violating the second law of thermodynamics if additional energy in the form of work or heat is provided to support the process.

In *compression heat pumps* we provide the work to reverse the heat flow, as shown in Fig. 1.12 (c). The amount of transferred energy, that is heat, is larger than the provided work, which means that the efficiency is above 100 %. In fact, the typical efficiency of a heat pump is many times larger than that of a furnace. Another advantage compared to the furnace is that it can be used both to replace the energy lost during the winter period and to dispose of the energy gained during the summer period. Compression heat pumps, which are capable of transferring heat only from the inside to the outside of the building and are therefore only used during the summer period, are generally referred to as air conditioners.

> **Info box**
>
> Heat pumps are more efficient than furnaces because they do not convert between forms of energy, but move energy between two environments.

The energy balance can also be achieved in conjunction to power stations, as shown in Fig. 1.12 (d). Power stations are industrial facilities that generate electrical energy. They basically consist of a furnace that converts chemical energy into internal energy, a heat engine that provides work when heat flows through it, and an electrical generator that converts work into electrical energy.

The by-product of the process is waste heat, which the power station can either release to the environment or supply to a building, as shown on the left side in Fig. 1.12 (d). In the latter case, the use of a power station is referred to as cogeneration or Combined Heat and Power (CHP).

The waste heat from the heat engine can also be used for cooling, as shown on the right side in Fig. 1.12 (d). The *absorption heat pump* absorbs some additional heat when the main heat flow passes through it. In this case, the use of a power station is referred to as trigeneration or Combined Cooling, Heat and Power (CCHP).

Note that in this section we have used work and electrical energy interchangeably for simplicity. In fact, work is converted into electrical energy by an electric generator, and electrical energy is converted into work by an electric motor. However, because the efficiency of both devices is virtually 100 %, this does not affect the results of our considerations. The exact functioning of heat pumps and heat engines is explained in more detail in Section 1.9.

1.9 Heat pumps and engines

In Section 1.8 we mentioned that the building energy balance can be achieved by using a compression heat pump, a device that in various forms becomes a standard building installation. Now we will take a closer look at the operating principle of the compression heat pump, which until the end of this section is simply referred to as a heat pump. For comparison, we will also look at the operating principle of its close sibling, the heat engine.

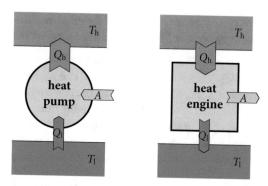

Figure 1.13: The principles of the heat pump (left) and the heat engine (right). When the heat pump is supplied with work, it transfers heat from an environment at lower temperature T_l to an environment at higher temperature T_h. When the heat engine is brought into contact with two environments at different temperatures T_h and T_l, it generates work.

1.9.1 Introduction

As we have already mentioned, heat pumps and heat engines exchange energy with the environment in the form of both heat and work.

Heat pump

The *heat pump* (Fig. 1.13, left) is a device that transfers heat from the environment at a lower temperature T_l to the environment at a higher temperature T_h. Because the direction of heat flow is opposite to the direction of spontaneous heat flow, work A must be provided, usually by an electric motor. The internal energy of the heat pump remains unchanged in the process, so we can write

$$A + Q_l = Q_h.$$

The heat pump can be used in heating mode to heat the environment at a higher temperature (heating system), or in cooling mode to cool the environment at a lower temperature (refrigerator, air conditioner). If we define the *coefficient of performance* COP as the ratio of what is gained, the heat transferred from or to the environment, to what is given, the work provided by an electric motor, we obtain different coefficients in heating and cooling mode.

In cooling mode, the gain corresponds to the heat extracted from the environment at a lower temperature Q_l, so the coefficient is

$$\mathrm{COP}_{\mathrm{cooling}} = \frac{Q_l}{A} = \frac{Q_l}{Q_h - Q_l}, \tag{1.33}$$

where Q_h is heat disposed to the environment at a higher temperature.

In heating mode, on the other hand, the gain corresponds to the heat provided to the environment at a higher temperature Q_h, so the coefficient is

$$\mathrm{COP}_{\mathrm{heating}} = \frac{Q_h}{A} = \frac{Q_h}{Q_h - Q_l}, \tag{1.34}$$

where Q_l is heat absorbed from the environment at a lower temperature. The coefficient of performance is larger in heating mode than in cooling mode.

Ideally, the heat could be transferred spontaneously from the environment at T_l to the environment at T_h, without the need for work, that is $A = 0\,\mathrm{J}$; In this case the coefficient would be infinite. The fact that it is impossible to design a machine that continuously transfers heat from the environment at lower temperature to the environment at a higher temperature without the input of work is a distinct form of the *second law of thermodynamics*.

Heat engine

The *heat engine* (Fig. 1.13, right) is a device that generates work from the heat. Two environments at a higher temperature T_h and at a lower temperature T_l must be provided. The heat flows spontaneously from the higher to the lower temperature, and this fact is used to generate work A in the process. The internal energy of the heat pump remains unchanged in the process, so we can write

$$Q_h = A + Q_l.$$

We define *efficiency* η as the ratio of what is gained, the work A, to what is given, the heat supplied by the environment at a higher temperature Q_h

$$\eta = \frac{A}{Q_h} = \frac{Q_h - Q_l}{Q_h} = 1 - \frac{Q_l}{Q_h}, \qquad (1.35)$$

where Q_l is the heat disposed to the environment at a lower temperature. Ideally, the work could be obtained without the intervention of the environment at T_l, that is $Q_l = 0\,\mathrm{J}$; In this case all of the heat provided Q_h would be converted into work and the efficiency would be $\eta = 1$. The fact that it is impossible to design a machine that continuously converts all the heat from the environment into work is a distinct form of the *second law of thermodynamics*. The efficiency is therefore always $\eta < 1$.

1.9.2 Gas processes

The intention of this book is to present a simple heat pump and heat engine using the ideal gas as a working medium. But before that, we have to familiarise ourselves with four ideal gas processes:

1. The *isothermal process* is the process in which the temperature of the gas T is constant. If we put constant temperature into the ideal gas law (1.12), we see that the product of pressure and volume is constant

$$p\,V = \frac{mRT}{M} = \text{constant}. \qquad (1.36)$$

2. The *isobaric process* is the process in which the pressure of the gas p is constant. If we put constant pressure into the ideal gas law (1.12), we see that the quotient of volume by temperature is constant

$$\frac{V}{T} = \frac{mR}{Mp} = \text{constant.} \tag{1.37}$$

3. The *isochoric process* is the process in which the volume of the gas V is constant. If we put constant volume into the ideal gas law (1.12), we see that the quotient of pressure by temperature is constant

$$\frac{p}{T} = \frac{mR}{MV} = \text{constant.} \tag{1.38}$$

4. The *adiabatic process* is the process in which the gas does not exchange heat Q with the environment. In this case all gas state variables (p, V, T) change and the relationship between them can be determined by the total differential of the ideal gas law (1.12)

$$p \, dV + V \, dp = \frac{m}{M} R \, dT.$$

If we put $dQ = 0$ in the first law of thermodynamics (1.23) and combine it with (1.25), we get

$$m \, c_V \, dT + p \, dV = 0.$$

Eliminating dT from these two expressions and using (1.27) we get

$$p \, dV + V \, dp = -\frac{R}{M c_V} p \, dV = -\frac{c_p - c_V}{c_V} p \, dV = (1 - \gamma) p \, dV,$$

where γ is the ratio of heat capacities

$$\gamma = \frac{c_p}{c_V}.$$

Ordering and integrating the expression we get

$$\frac{dp}{p} + \gamma \frac{dV}{V} = 0,$$

$$\ln p + \gamma \ln V = \text{constant,}$$

$$p \, V^\gamma = \text{constant.} \tag{1.39}$$

With the help of the ideal gas law (1.12), we can also write the relationship between temperature and volume as

$$T \, V^{\gamma-1} = \text{constant.} \tag{1.40}$$

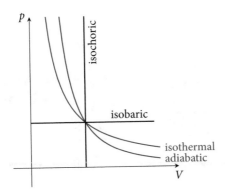

Figure 1.14: The curves for four ideal gas processes in the PV diagram. Note that the adiabatic curve is steeper than the isothermal curve.

The gas processes are normally examined in the pressure versus volume plot, called PV diagram. This is because in this case the area under the curve represents the work done by the gas (1.14). The PV diagram for all four processes is shown in Fig. 1.14. Because $c_p > c_V$ (1.27), $\gamma > 1$ and the adiabatic curve is steeper than the isothermal curve.

At the end of this section, we will derive the expression for the modulus of compression K for the adiabatic processes that we will need in Section 6.1.1 on page 187. If we take the definition of the modulus as

$$\Delta p = -K\frac{\Delta V}{V},$$

$$\frac{\Delta p}{\Delta V} = -\frac{K}{V},$$

write it in differential form and use $pV^\gamma = C$ (1.39), we get

$$\frac{K}{V} = -\frac{dp}{dV} = -\frac{d}{dV}(CV^{-\gamma}) = \gamma CV^{-\gamma-1} = \gamma\frac{p}{V}$$

$$\implies K = \gamma\,p. \tag{1.41}$$

1.9.3 Carnot cycle

Finally, we will demonstrate a simple heat pump and a simple heat engine using the ideal gas as a working medium. The gas is located in a cylinder fitted with a movable piston at one end. The process will consist of two isothermal and two adiabatic processes called the Carnot cycle. We have chosen this process because it is the most efficient. Note that proving this fact would go beyond the scope of this book.

> **Info box**
>
> All heat pumps and heat engines are based on cyclical thermodynamic processes.

Heat pump

The heat pump cycle is shown in Fig. 1.15 on the left, with the graphic representation above and the PV diagram below. The four processes are:

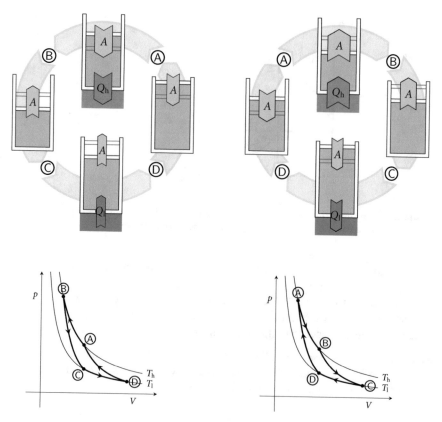

Figure 1.15: Heat pump (left) and heat engine (right) using Carnot cycle. The figures above are graphical representation, while the plots below are corresponding PV diagrams. Note that the heat pump and heat engine use the same cycle in the opposite direction.

1. Process Ⓐ to Ⓑ: The gas is in thermal contact with the environment at T_h. Because we compress the gas isothermally, its internal energy remains constant. Due to the first law of thermodynamics, this means that the provided work is completely converted into heat, which is released into the environment. This process can only take place up to a certain maximum pressure or minimum volume. Therefore we have to bring the gas into contact with the environment at T_l to absorb more energy and then return to point Ⓐ.

2. Process Ⓑ to Ⓒ: Before we bring the gas into contact with the environment at T_l, we have to cool it down. So we let it expand adiabatically and harness the work that is provided in this way. The process ends when the gas reaches the temperature T_l.

3. Process Ⓒ to Ⓓ: The gas is in thermal contact with the environment at T_l. We let the gas expand isothermally so that its internal energy remains constant. Due to first law of thermodynamics, this means that the heat that the gas absorbs from the environment is completely converted into the work, which we harness. This process can only take place up to a certain minimum pressure or maximum volume.

4. Process Ⓓ to Ⓐ: Before we bring the gas into contact with the environment at T_h, we have to warm it up. So we compress it adiabatically until the gas reaches temperature T_h. The cycle is complete.

The PV diagram allows us to estimate the work we have invested and harnessed. The work we have invested is the area under the Ⓓ-Ⓐ-Ⓑ curve, and the work we have harnessed is the area under the Ⓑ-Ⓒ-Ⓓ curve. So the work we have harnessed is smaller than the work we have invested, which is the condition to transfer the heat from the environment at T_l to the environment at T_h.

Heat engine

The heat engine cycle is shown in Fig. 1.15 on the right, with the graphic representation above and the PV diagram below. The four processes are:

1. Process Ⓐ to Ⓑ: The gas is in thermal contact with the environment at T_h. Because we let the gas expand isothermally, its internal energy remains constant. Due to the first law of thermodynamics, this means that all the heat that the gas absorbs from the environment is completely converted into the work that we harness. This process can only take place up to a certain maximum volume or minimum pressure, and we must return to point Ⓐ to gain more work. Of course it makes no sense to use the same path, because then we would be spending all the work we just harnessed. However, it turns out that if we compress gas at a lower temperature, we will need less work to get to that point. So we have to bring the gas into contact with the environment at T_l first.

2. Process Ⓑ to Ⓒ: Before we bring the gas into contact with the environment at T_l, we have to cool it down. So we let it to expand adiabatically and harness the work that is provided in this way. The process ends when the gas reaches the temperature T_l.

3. Process Ⓒ to Ⓓ: The gas is in thermal contact with the environment at T_l. We compress the gas isothermally so that its internal energy remains constant. Due to the first law of thermodynamics, this means that the provided work is completely converted into heat, which is released into the environment. This process can only take place up to a certain maximum pressure or minimum volume.

4. Process Ⓓ to Ⓐ: Before we bring the gas into contact with the environment at T_h, we have to warm it up. So we compress it adiabatically, until the gas reaches temperature T_h. The cycle is complete.

The PV diagram allows us to estimate the work we have invested and harnessed. The work we have invested is the area under the Ⓒ-Ⓓ-Ⓐ curve, and the work we have harnessed is the area under the Ⓐ-Ⓑ-Ⓒ curve. So the work we have harnessed is larger than the work we have invested, which has been achieved by transferring heat from the environment at T_h to the environment at T_l.

Calculation

We will now calculate the heat that is absorbed or released by the gas during the isothermal processes. As already mentioned, due to the first law of thermodynamics, heat is equal to work

$$Q = A = \int_{V_1}^{V_2} p \, dV = \frac{mRT}{M} \int_{V_1}^{V_2} \frac{dV}{V} = \frac{mRT}{M} \ln\left(\frac{V_2}{V_1}\right),$$

where we have used the ideal gas law (1.12) and the expression for isothermal processes (1.36). Using the formula for Ⓐ-Ⓑ and Ⓓ-Ⓒ processes we obtain

$$|Q_h| = \pm\frac{mRT_h}{M} \ln\left(\frac{V_B}{V_A}\right), \qquad |Q_l| = \pm\frac{mRT_l}{M} \ln\left(\frac{V_C}{V_D}\right),$$

where the signs '+' and '-' stand for the heat pump and the heat engine respectively. On the other hand, we can connect volume pairs by the expression for adiabatic processes (1.40)

$$T_h V_B^{\gamma-1} = T_l V_C^{\gamma-1}, \qquad T_h V_A^{\gamma-1} = T_l V_D^{\gamma-1}.$$

Dividing the two equations we get

$$\left(\frac{V_B}{V_A}\right)^{\gamma-1} = \left(\frac{V_C}{V_D}\right)^{\gamma-1} \implies \frac{V_B}{V_A} = \frac{V_C}{V_D}.$$

Now we can calculate the ratio of the heats as

$$\frac{|Q_h|}{|Q_l|} = \frac{T_h \ln\left(\frac{V_B}{V_A}\right)}{T_l \ln\left(\frac{V_C}{V_D}\right)} = \frac{T_h}{T_l}. \tag{1.42}$$

If we insert (1.42) in (1.33) and (1.34), we get for coefficients of performance

$$\text{COP}_{\text{cooling}} = \frac{T_l}{T_h - T_l}, \tag{1.43}$$

$$\text{COP}_{\text{heating}} = \frac{T_h}{T_h - T_l}. \tag{1.44}$$

> **Info box**
>
> The coefficient of performance of the heat pump depends on the temperature difference between two environments.

Note that the coefficients depend on the temperature of the environments: If the temperature difference is smaller, the coefficient of performance is larger.

If we put (1.42) into (1.35) we get for efficiency

$$\eta = 1 - \frac{T_l}{T_h}. \tag{1.45}$$

Note that the efficiency depends on the temperature of the environments: If the temperature ratio is larger, the efficiency is also larger.

We have already pointed out the advantage of the Carnot cycle. Unfortunately, it is not technically feasible to design the heat pump and the heat engine with this cycle. Nevertheless, it helps us to reveal the operating principles of heat pumps and heat engines and provide us with the top theoretical values of the coefficient of performance and the efficiency.

Problems

1.1 When the copper bar length is measured using a steel tape at $-10\,°C$, the measured value is 2000.0 mm. What is the measurement value at 30 °C? Take linear expansion coefficients of copper and steel to be $1.7 \times 10^{-5}\,K^{-1}$ and $1.2 \times 10^{-5}\,K^{-1}$, respectively. (2000.4 mm)

1.2 The volume of alcohol in a thermometer glass bulb at 0 °C is 200 mm³. The bulb is attached to a narrow glass tube of diameter 0.50 mm. Calculate the 'length' of degree Celsius on the glass tube. Assume that the glass does not expand. Take the cubic expansion coefficient of alcohol to be $1.1 \times 10^{-3}\,K^{-1}$. (1.1 mm)

1.3 We want to put 2.0 kg of oxygen gas at 20 °C into a gas cylinder of volume 100 L. What minimum pressure should the gas cylinder be able to endure? Take the molar mass of oxygen to be 0.032 kg/mol. (1.5×10^6 Pa)

1.4 Calculate the density of dry air at pressure 1.00 bar and (a) at 10.0 °C and (b) at 20.0 °C. Take the molar mass of dry air to be 29.0 g/mol. ($1.23\,kg/m^3$, $1.19\,kg/m^3$)

1.5 The gas mixture of mass 10.0 kg and molar mass 0.026 kg/mol contains oxygen of mass 2.0 kg and molar mass 0.032 kg/mol. If the pressure of the gas mixture amounts to 2.5 bar, what is the partial pressure of the oxygen? (0.41 bar)

1.6 Air at atmospheric pressure 1.013×10^5 Pa and at 10 °C is in a room of volume 10 m³. Calculate the heat that should be provided to warm up the air to 30 °C if (a) the room is well sealed and if (b) the room is leaky. Take the molar mass of air to be 29 g/mol and the specific heat capacity at constant volume to be 720 J/(kg K). (180 kJ, 250 kJ)

1.7 A stainless steel vessel of mass 600 g contains 2.0 L of liquid water, both at a temperature of 20 °C. A 1.0 kW heater is inserted into the vessel and turned on. Calculate the time needed to bring the vessel and water to the temperature of boiling water. Take specific heat capacities of stainless steel and water to be 600 J/(kg K) and 4200 J/(kg K), respectively. (12 min)

1.8 A glass sphere of radius 2.5 cm at 150 °C is tossed into water of mass 2.0 kg at 18 °C. Calculate the equilibrium temperature. Take the density of glass to be $2.5 \times 10^3\,kg/m^3$, as well as specific heat capacities of water and glass to be 4200 J/(kg K) and 800 J/(kg K), respectively. (20 °C)

1.9 An aluminium cube of mass 1.0 kg at 600 °C is tossed into water of mass 1.0 kg at 20 °C. Calculate the mass of evaporated water. Take the specific heat of vaporisation of water to be 2.26×10^6 J/kg, as well as specific heat capacities of water and aluminium to be 4200 J/(kg K) and 900 J/(kg K), respectively. (50 g)

1.10 Ice of mass 500 g at −10 °C and steel of mass 1.0 kg at 500 °C are placed into a well-insulated steel vessel of mass 450 g at 25 °C. Calculate the equilibrium temperature. Take the specific heat of fusion of water to be 336 kJ/kg, as well as the specific heat capacities of water, ice and steel to be 4200 J/(kg K), 2100 J/(kg K) and 470 J/(kg K), respectively. (22 °C)

1.11 A well-insulated vessel contains 2.0 kg of liquid water at 0 °C. Water is frozen by pumping out the water vapour until all liquid either evaporates or freezes. Calculate the mass of the ice obtained. Take the specific heat of fusion to be 336 kJ/kg and the specific heat of evaporation at 0 °C to be 2.5 MJ/kg. (1.8 kg)

1.12 The room of rectangular floor plan with dimensions 10.0 m × 20.0 m and height 3.0 m contains dry air of density 1.165 kg/m³ at temperature 30 °C. In the room we place a bucket with 12 L of liquid water at 5 °C. After a while, the temperatures of air and water will equalize at 20 °C. Calculate the mass of the evaporated water. Take the specific heat capacity of dry air to be 1005 J/(kg K), the specific heat capacity of liquid water to be 4200 J/(kg K) and the specific heat of vaporisation of water at 20 °C to be 2450 kJ/kg. (2.6 kg)

1.13 An average adult human being consumes most energy on heat loss, through radiation, convection, perspiration, respiration and conduction (listed in order of importance). Calculate the power due to perspiration and respiration under assumptions:

- For perspiration and respiration, an average person consumes 0.8 L of liquid water at 20 °C per day, with the water being converted into vapour at 37 °C. The specific heat capacity of water is 4200 J/(kg K) and the specific heat of vaporisation of water at 37 °C is 2410 kJ/kg.

- An average person inhales and exhales 11 m³ of air in the room, with the air being heated from 20 °C to 37 °C. The density of inhaled air is 1.2 kg/m³ and the specific heat capacity is 1010 J/kg.

We will calculate other heat losses in problem 2.6 in Chapter 2. (26 W)

2 Heat transfer

2.1 Introduction

In this chapter, we will look more closely at mechanisms that facilitate heat transfer. Heat can be transferred by three basic mechanisms, which are

1. *conduction,*
2. *convection* and
3. *radiation.*

In short time periods, the amount of transferred heat is practically always proportional to time. Heat transferred in 1 min will be 60 times larger than heat transferred in 1 s. It is therefore convenient to define the *heat flow rate* with the unit *watt* Φ (W) as the quotient of transferred heat by time:

$$\Phi = \frac{dQ}{dt}. \tag{2.1}$$

The watt therefore is equal to J/s. For a stationary situation, that is, a time-independent heat flow, a nondifferential form can be used:

$$\Phi = \frac{Q}{t}. \tag{2.2}$$

In most practical situations, the heat flow rate is proportional to the area, so it is also convenient to define the *density of heat flow rate* q (W/m^2) as

$$\Phi = A\,q. \tag{2.3}$$

> **Info box**
>
> The transfer of heat is best described by the density of heat flow rate.

2.2 Conduction

Heat *conduction* is a mechanism of heat transfer facilitated by microscopic particles without bulk movement of particles. This mechanism is typical for solids (see Section 1.1 on page 3) because most particles are moving only around a fixed equilibrium position (oscillation), although some (electrons) may also move randomly through the material (*diffusion*). Heat

> **Info box**
> Conduction is facilitated by stationary particles, so it primarily occurs in the solid state.

> **Info box**
> Equilibrium stands for temperatures independent of space and time. Steady heat transfer stands for the space-dependent and time-independent temperatures. Steady heat transfer is studied due to its simplicity.

> **Info box**
> If there is a spatial difference in temperature, the conduction—heat diffusion—from the region with the higher temperature to the region with the lower temperature sets in.

transfer is a consequence of *collisions between particles* and *particle diffusion*. Note that in terms of energy, the process is diffusive, so conduction can be also called *heat diffusion*.

2.2.1 Fourier's law

We start our study of conduction with *steady heat transfer*. Steady implies that we allow for temperature variations in space but no variations in time.

Let's observe heat transfer through a solid slab whose opposite faces are at different temperatures; that is, the warmer face is at higher temperature T_h, and the colder face is at lower temperature T_l (Fig. 2.1). The slab's cross-sectional area is A, and the thickness is d.

It can be shown experimentally that the heat flow rate is always

- proportional to cross-sectional area A,
- inversely proportional to thickness d and
- proportional to *temperature difference* $\Delta T = T_h - T_l$.

We can join these statements into the expression

$$\Phi = \lambda \frac{A \Delta T}{d}. \tag{2.4}$$

This statement is called *Fourier's law* or the *law of thermal conduction*. The coefficient of proportionality λ ($W/(m\,K)$) is called *thermal conductivity*.

Thermal conduction also depends on the slab substance. Thermal conductivity is thus material dependent and has to be determined experimentally. Values for a few typical building materials are presented in Table A.3 on page 277.

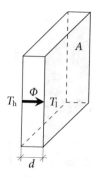

Figure 2.1: Heat transfer (conduction) through a solid slab whose opposite faces are at different temperatures, $T_h > T_l$. The slab's cross-sectional area is A, and the thickness is d.

Figure 2.2: Thermal insulators: extruded polystyrene (left) and mineral wool (right). The primary ingredient of insulators is air (see Section 2.3).

Thermal insulators are materials that hinder heat transfer (Fig. 2.2). Note that good thermal insulators have *low* thermal conductivity.

Because the temperature difference is the same for both temperature scales (1.2), and degrees Celsius is used more commonly in engineering, we will rewrite Fourier's law as

> **Info box**
>
> Thermal insulators, that is, materials that hinder heat transfer, have low thermal conductivity.

$$\Phi = \lambda \frac{A \Delta \theta}{d}. \tag{2.5}$$

In terms of density of heat flow rate (2.3), Fourier's law is transformed to

$$q = \lambda \frac{\Delta \theta}{d}. \tag{2.6}$$

Usually, the layer in the building component is characterised by *thermal resistance* R (m^2 K/W), that is, the quotient of thickness by thermal conductivity

$$R = \frac{d}{\lambda}, \tag{2.7}$$

or, alternatively, though rarely, by the inverse of thermal resistance, that is, the *coefficient of heat transfer* k (W/(m^2 K)) as

$$k = \frac{1}{R} = \frac{\lambda}{d}. \tag{2.8}$$

In terms of thermal resistance, Fourier's law is therefore

$$\Phi = \frac{A \Delta \theta}{R}, \tag{2.9}$$

$$q = \frac{\Delta \theta}{R}. \tag{2.10}$$

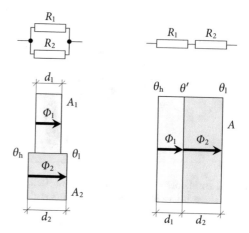

Figure 2.3: Conduction through multiple layers in building components when they are positioned in parallel (left) or serial (right). Equivalent electrical circuits are drawn above.

2.2.2 Thermal resistance of multiple layers

We will now attend to the problem of heat transfer through multiple layers in building components when they are in parallel or serial positions (Fig. 2.3).

For *parallel layers* (Fig. 2.3, left), both layers have temperature θ_h at the left face and temperature θ_l at the right face. Taking this into consideration, we can write

$$\Phi_1 = \frac{A_1(\theta_h - \theta_l)}{R_1}, \quad \Phi_2 = \frac{A_2(\theta_h - \theta_l)}{R_2}.$$

The effective heat flow rate from the left side to the right side is simply the sum of the heat flow rates through the individual layers:

$$\Phi = \Phi_1 + \Phi_2 = \left(\frac{A_1}{R_1} + \frac{A_2}{R_2}\right)(\theta_h - \theta_l) \equiv \frac{A_{eq}}{R_{eq}}(\theta_h - \theta_l)$$

$$\implies \frac{A_{eq}}{R_{eq}} = \frac{A_1}{R_1} + \frac{A_2}{R_2}.$$

We can generalise this result for an arbitrary number n of layers as

$$\frac{A_{eq}}{R_{eq}} = \sum_{i=1}^{n} \frac{A_i}{R_i}. \tag{2.11}$$

For *serial layers* (Fig. 2.3, right), both layers have the same cross-sectional area $A_1 = A_2 = A$. There is also another temperature involved, that is, the temperature at the boundary of two layers θ', which is usually called the *interface temperature*. Taking this into account, we can write

$$\Phi_1 = \frac{A(\theta_h - \theta')}{R_1}, \qquad \Phi_2 = \frac{A(\theta' - \theta_l)}{R_2}.$$

What is the relationship between the heat flow rates in both layers? If the first heat flow rate is larger than the second, $\Phi_1 > \Phi_2$, then more heat will enter the boundary of the two layers than will leave it; therefore, temperature θ' will increase. On the other hand, if the first heat flow rate is smaller than the second, $\Phi_1 < \Phi_2$, then less heat will enter the boundary of the two layers than will leave it; therefore, temperature θ' will decrease. We conclude that in order to have steady heat transfer, that is, constant temperature θ', both flow rates must be equal as in

$$\Phi_1 = \Phi_2.$$

Using that, we can readily calculate

$$\theta' = \frac{\theta_1 R_1 + \theta_h R_2}{R_1 + R_2}.$$

Putting θ' in the equation for Φ_1 or Φ_2, we obtain

$$\Phi = \Phi_1 = \Phi_2 = \frac{A}{R_1 + R_2}(\theta_h - \theta_1) \equiv \frac{A_{eq}}{R_{eq}}(\theta_h - \theta_1)$$

$$\implies R_{eq} = R_1 + R_2.$$

Here we took into account that $A_{eq} = A$.

For an arbitrary number n of layers

$$\Phi = A\frac{\theta_h - \theta'_1}{R_1} = A\frac{\theta'_1 - \theta'_2}{R_2} = A\frac{\theta'_2 - \theta'_3}{R_3} = \cdots = A\frac{\theta'_{n-1} - \theta_1}{R_n}. \qquad (2.12)$$

Using easily provable rule

$$f = \frac{a_1}{b_1} = \frac{a_2}{b_2} = \frac{a_3}{b_3} = \cdots = \frac{a_n}{b_n} \implies f = \frac{a_1 + a_2 + a_3 + \cdots + a_n}{b_1 + b_2 + b_3 + \cdots + b_n}, \qquad (2.13)$$

we similarly get

$$\Phi = A\frac{\theta_h - \theta_1}{R_1 + R_2 + R_3 + \cdots + R_n},$$

which leads to the generalised form of

$$R_{eq} = \sum_{i=1}^{n} R_i. \qquad (2.14)$$

It is always instructive to point to the similarity between heat conduction and electrical current conduction for electrical circuits, which is shown in Fig. 2.3 above. In the parallel case, heat splits and flows partially through the upper layer and partially through the lower layer. In the serial case, heat flows through both layers. Similarly, in the parallel case, electrical current splits and flows partially through the upper and partially through the lower resistor; in the serial case, electrical current flows through both resistors.

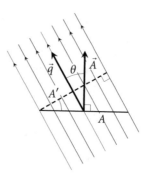

Figure 2.4: Nonperpendicular heat flow through a surface of area A. The scalar density of heat flow rate is defined in terms of the surface perpendicular to heat flow of area A', so angle of flow incidence θ has to be taken into account.

2.2.3 Multidimensional conduction

If heat does not flow perpendicularly to the surface, angle of incidence θ has to be taken into account (Fig. 2.4). Equation (2.3) is defined in terms of the surface perpendicular to heat flow of area A', so heat flow in the slanted case amounts to

$$\Phi = A'q = A\cos\theta\, q. \tag{2.15}$$

We see that the heat flow rate can differ for the same density of heat flow rate and the same surface area but a different angle of flow incidence. In Section 2.4, we will use that fact to explain different temperatures on the Earth's surface.

We can define heat flow rate as a vector

$$\vec{q} = q_x\vec{i} + q_y\vec{j} + q_z\vec{k} \tag{2.16}$$

and mathematically describe the surface in terms of a vector $\vec{A}$ perpendicular to the surface with the length corresponding to the surface area. In this case, the heat flow rate can be rewritten in terms of a scalar product of those two vectors

$$\Phi = \vec{A}\cdot\vec{q}. \tag{2.17}$$

In one dimension, for thin layers, the thickness and temperature difference in (2.6) tend to zero, $d \rightarrow dx, \Delta\theta \rightarrow d\theta$, leading to the differential version

$$q = -\lambda\frac{d\theta}{dx}, \tag{2.18}$$

where we have taken into account that when temperature increases in the $+x$ direction, heat flows in the opposite $-x$ direction and vice versa.

In more complex situations, however, temperature is a function of all three coordinates, $\theta(x, y, z)$, and heat flows in all three directions. We must therefore write Fourier's law for each of the dimensions

$$q_x = -\lambda \frac{\partial \theta}{\partial x}, \quad q_y = -\lambda \frac{\partial \theta}{\partial y}, \quad q_z = -\lambda \frac{\partial \theta}{\partial z},$$

where the symbol ∂ denotes a partial derivative. Using the density of heat flow rate vector (2.16) and *nabla operator*

$$\vec{\nabla} = \frac{\partial}{\partial x}\vec{i} + \frac{\partial}{\partial y}\vec{j} + \frac{\partial}{\partial z}\vec{k}, \tag{2.19}$$

we can transform the Fourier's law to a three-dimensional form of

$$\vec{q} = -\lambda \left(\frac{\partial \theta}{\partial x}\vec{i} + \frac{\partial \theta}{\partial y}\vec{j} + \frac{\partial \theta}{\partial z}\vec{k} \right),$$

$$\vec{q} = -\lambda \vec{\nabla} \theta. \tag{2.20}$$

The density of heat flow rate is therefore the product of the thermal conductivity and negative *gradient* of temperature.

The gradient operator is the generalisation of the derivative operator in multiple dimensions. If the value of a certain scalar quantity is specified by three-dimensional function $f(x, y, z)$, the gradient at an arbitrary position returns the vector with two properties:

1. The vector is directed along the increase of the function.

2. The vector length is proportional to the slope of the increase.

Figure 2.5 presents an example of a gradient for a two-dimensional function.

The scalar quantity of interest in our case is temperature, whereas the vector is the density of the heat flow rate. Because the heat flow rate is directed along the decrease of temperature, whereas the gradient is directed along the increase of the temperature, the minus sign in (2.20) is required in order to reverse the direction of the gradient vector.

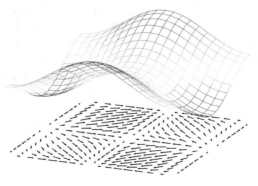

Figure 2.5: An illustrative two-dimensional function $f(x, y) = \cos x - \cos y$. The gradient (vector) is directed towards the increase of the function, and its length is proportional to the slope of the function.

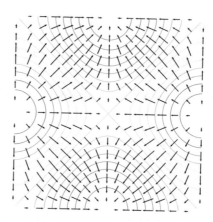

Figure 2.6: Contour lines of an illustrative two-dimensional function $f(x, y) = \cos x - \cos y$. The gradient vector is perpendicular to the contour lines, and its length is proportional to the density of the contour lines.

For any multidimensional function, it is also possible to define contour lines, which are curves that connect points in space with the same value. The gradient vector has two properties:

1. The vector is perpendicular to the contour lines.

2. The vector length is proportional to the density of the contour lines.

Figure 2.6 presents an example of contour lines and gradient for a two-dimensional function. Here we are interested in the isothermal lines, which are curves that connect points with the same temperature. The density of heat flow rate vector is therefore always perpendicular to the isothermal lines.

> **Info box**
>
> The density of heat flow rate vector is perpendicular to the isothermal lines.

A practical example of two-dimensional conduction, temperatures and contours for a simple geometric thermal bridge is shown later in Fig. 3.13 on page 90.

2.2.4 Dynamic conduction

Until now, we have studied *steady heat transfer* (conduction), that is, situations with time-independent temperatures that disregard the accumulation of energy in building components. Because the temperature and internal energy of the building component are constants, the heat flow rate entering the building component on one side is equal to the heat flow rate leaving the component on the other side (Fig. 2.7, left).

> **Info box**
>
> Dynamic heat transfer conditions occur for time-dependent temperatures and are used for studying more complex processes.

Now we will take a closer look at *dynamic heat transfer* (conduction). Because the heat flow rate entering the building component on one side differs from the heat flow rate leaving the component on the other side, the building component's internal energy and temperature must change (Fig. 2.7, middle and right).

In order to describe temperature change, we observe a small fragment of the building component with dimensions $\Delta x \times \Delta y \times \Delta z$ (Fig. 2.8). In a

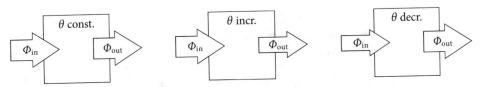

Figure 2.7: Difference between steady conduction (left) and dynamic conduction (middle, right). When heat flow rate entering building component differs from the one leaving, temperature of building component changes.

dynamic situation, the heat flow rate that enters fragment, Φ_{in}, is different from the heat flow rate that leaves it, Φ_{out}. The difference of heat flow rates corresponds to the net heat transferred to/from the fragment in a given time period (2.1):

$$\Phi_{net} = \Phi_{in} - \Phi_{out} = \frac{dQ}{dt}.$$

Because the heat flow is not necessarily directed along one of the coordinate axes, we have to decompose it into components

$$\vec{\Phi} = \Phi_x \vec{i} + \Phi_y \vec{j} + \Phi_z \vec{k}.$$

As shown in Fig. 2.8, x, y and z components of heat flow rate Φ_x, Φ_y and Φ_z enter and leave the fragment along x, y and z axes, respectively. For simplicity, we will assume that $\Phi_{in} > \Phi_{out}$ and consequently conclude that the net heat flow rate to the fragment increases its temperature (1.21):

$$[\Phi_x(x)+\Phi_y(y)+\Phi_z(z)]-[\Phi_x(x+\Delta x)+\Phi_y(y+\Delta y)+\Phi_z(z+\Delta z)] = mc\frac{\partial\theta}{\partial t}.$$

The mass of the fragment can be written in terms of its density and volume as

$$m = \rho V = \rho \, \Delta x \Delta y \Delta z.$$

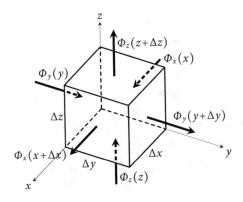

Figure 2.8: A small fragment of the building component with dimensions $\Delta x \times \Delta y \times \Delta z$. In a dynamic situation, the heat flow rate that enters the fragment is different from the heat flow rate that leaves it, which leads to the temperature change.

Because the heat flow rate is a smooth function, the exiting heat flow rate components can be related to the entering heat flow rate components using Taylor series

$$\Phi_x(x+\Delta x) = \Phi_x(x) + \frac{\Delta x}{1!}\frac{\partial \Phi_x}{\partial x} + \frac{\Delta x^2}{2!}\frac{\partial^2 \Phi_x}{\partial x^2} + \cdots \approx \Phi_x(x) + \Delta x\frac{\partial \Phi_x}{\partial x},$$

$$\Phi_y(y+\Delta y) = \Phi_y(y) + \frac{\Delta y}{1!}\frac{\partial \Phi_y}{\partial y} + \frac{\Delta y^2}{2!}\frac{\partial^2 \Phi_y}{\partial y^2} + \cdots \approx \Phi_y(y) + \Delta y\frac{\partial \Phi_y}{\partial y},$$

$$\Phi_z(z+\Delta z) = \Phi_z(z) + \frac{\Delta z}{1!}\frac{\partial \Phi_z}{\partial z} + \frac{\Delta z^2}{2!}\frac{\partial^2 \Phi_z}{\partial z^2} + \cdots \approx \Phi_z(z) + \Delta z\frac{\partial \Phi_z}{\partial z},$$

where we have neglected the higher order contributions. By using the preceding equations, we obtain

$$-\Delta x\frac{\partial \Phi_x}{\partial x} - \Delta y\frac{\partial \Phi_y}{\partial y} - \Delta z\frac{\partial \Phi_z}{\partial z} = \rho c\, \Delta x\Delta y\Delta z\,\frac{\partial \theta}{\partial t}.$$

The density of heat flow rate is the heat flow rate divided by the corresponding cross-sectional area

$$q_x = \frac{\Phi_x}{\Delta y\Delta z},\ q_y = \frac{\Phi_y}{\Delta x\Delta z},\ q_z = \frac{\Phi_z}{\Delta x\Delta y},$$

leading to

$$\frac{\partial q_x}{\partial x} + \frac{\partial q_y}{\partial y} + \frac{\partial q_z}{\partial z} = -\rho c\frac{\partial \theta}{\partial t},$$

$$\vec{\nabla}\cdot\vec{q} = -\rho c\frac{\partial \theta}{\partial t}. \tag{2.21}$$

Info box

Materials that damp temporal temperature variations more intensely have larger volumetric heat capacities.

Here we used the definition of the heat flow rate as a vector (2.16) and nabla operator (2.19). This is the *heat continuity equation* in absence of heat sources, which states that the *divergence* of the density of heat flow rate is the product of the volumetric heat capacity (1.22) and temperature change. Note that materials with larger volumetric heat capacities will have smaller temperature changes for the same heat flow rate; therefore, they will damp the temporal temperature variations more intensely.

The divergence operator is somewhat similar to a gradient. It probes the values of vectors in the vicinity of the arbitrary position. If the magnitudes of vectors pointing out of the infinitesimal environment are larger than the magnitudes of vectors pointing towards the infinitesimal environment, divergence is positive, and vice versa.

Joining the heat continuity equation (2.21) and the three-dimensional Fourier's law (2.20), we obtain the following *heat diffusion equation*:

$$\vec{\nabla}\cdot(\lambda\vec{\nabla}\theta) = \rho c\frac{\partial \theta}{\partial t}. \tag{2.22}$$

The heat diffusion equation is a partial differential equation of the second order that describes the spatial and temporal variations of temperature θ.

It is also used to study heat conduction, where the equation is first solved to determine temperatures, after which the density of heat flow rate is obtained using (2.20). Finally, the heat flow rate can be obtained by integrating the density of heat flow rate over surface A using the differential version of the expression (2.17):

$$\Phi = \int_A \vec{q} \cdot \mathrm{d}\vec{A}.$$

Note that, in general, thermal conductivity depends on spatial coordinates, as well as temperature and mass concentration of water w (see Section 4.4 on page 129), $\lambda = f(x, y, z, \theta, w)$; therefore, solving differential equations for real problems can be extremely complicated. Usually, the solution is found by the splitting building component into parts with constant thermal conductivity, in which case, (2.22) is simplified into the form

$$\frac{\partial \theta}{\partial t} = \left(\frac{\partial^2 \theta}{\partial x^2} + \frac{\partial^2 \theta}{\partial y^2} + \frac{\partial^2 \theta}{\partial z^2} \right),$$

$$\frac{\partial \theta}{\partial t} = a \nabla^2 \theta, \tag{2.23}$$

where *thermal diffusivity* a (m^2/s) is the quotient of thermal conductivity by volumetric heat capacity (1.22):

$$a = \frac{\lambda}{\rho c}. \tag{2.24}$$

Values for a few typical building materials are presented in Table 2.1 on page 44. Operator

$$\nabla^2 = \frac{\partial^2}{\partial x^2} + \frac{\partial^2}{\partial y^2} + \frac{\partial^2}{\partial z^2} \tag{2.25}$$

is called the *Laplace operator*. For a proximity of an arbitrary position, the Laplace operator of the function presents the rate at which the average value of the function deviates from the central value as the distance increases.

Because differential equations (2.22) and (2.23) include a time derivative, we also need the following to get the solution:

- *Boundary conditions*, value of the temperature or density of heat flow rate at the boundary of the observed system.

- *Initial conditions*, value of the temperature within the observed system at the initial moment.

Practically all real problems are so complex that the solutions to (2.22) and (2.23) must be found numerically. However, there are a few idealised theoretical models that have analytical solutions and are instructive enough to be presented here. Because this is beyond the scope of this book, the mathematical procedure for solving differential equations will be skipped, and only the solution to the model will be revealed. All models of interest are one-dimensional, so we will drop the x and y dependence and look for $\theta(z, t)$ as the solution of

$$\frac{\partial \theta}{\partial t} = a \frac{\partial^2 \theta}{\partial z^2}. \tag{2.26}$$

2.2.5 Contact of two materials at different temperatures

In the first theoretical model, we assume that the half-space ($z \leq 0$) is occupied by one material of known thermal conductivity λ_1, density ρ_1 and specific heat capacity c_1 at initial temperature θ_1', and the other half-space ($z \geq 0$) is occupied by another material of known thermal conductivity λ_2, density ρ_2 and specific heat capacity c_2 at initial temperature θ_2'. At the initial moment, we put those materials into contact, which leads to a heat flow from the material at the higher temperature to the material at the lower temperature. The obvious boundary conditions are that the temperatures of the both materials on the common boundary must be equal as

$$\theta_1(z = 0) = \theta_2(z = 0),$$

and the density of heat flow rate leaving one material must be equal to the density of heat flow rate entering the other (2.18):

$$\lambda_1 \left. \frac{d\theta_1}{dz} \right|_{z=0} = \lambda_2 \left. \frac{d\theta_2}{dz} \right|_{z=0}.$$

The solution of equation (2.26) is in the form

$$\theta_1(z, t) = \theta_0 + (\theta_1' - \theta_0) \operatorname{erf}\left(\frac{z}{2\sqrt{a_1 t}}\right),$$

$$\theta_2(z, t) = \theta_0 + (\theta_2' - \theta_0) \operatorname{erf}\left(\frac{z}{2\sqrt{a_2 t}}\right),$$

where

$$\operatorname{erf}(z) = \frac{2}{\sqrt{\pi}} \int_0^z e^{-t^2} dt$$

is the error function. The contact temperature (temperature at the boundary of two materials) is

$$\theta_0 = \frac{b_1 \theta_1' + b_2 \theta_2'}{b_1 + b_2}, \tag{2.27}$$

where

$$b = \sqrt{\lambda \rho c} \tag{2.28}$$

is *thermal effusivity* b ($\mathrm{W\,s^{\frac{1}{2}}/(m^2\,K)}$). Note that the contact temperature is closer to the temperature of the material with the larger thermal effusivity,

---

Info box

The contact temperature is closer to the temperature of the material with the larger thermal effusivity.

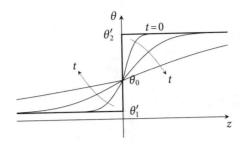

Figure 2.9: Spatial distribution of temperature for the contact of two materials at different temperatures. Initially, both materials have constant temperatures θ_1' and θ_2'. Subsequently, the temperatures of the materials near the contact gradually change towards the contact temperature θ_0.

which means that the material with the *larger thermal effusivity* resists the temperature change *at the surface* more strongly.

In Fig. 2.9, spatial distribution of temperature is presented after two materials with different temperatures come into contact. Constant contact temperature θ_0 is immediately established, whereas the temperature of the bulk of the material gradually drifts towards the contact temperature. A smaller thermal diffusivity inhibits the temperature equalisation, which means that the material with the *smaller thermal diffusivity* resists the temperature change *within* more strongly.

A practical application of this theoretical model is experienced when a bare foot comes into contact with a room floor. The temperature of the foot is about 36 °C, whereas, in general, the temperature of the floor is equal to the temperature of the room, regardless of the material. However, ceramic tiles have a much larger thermal effusivity than timber parquet (Table 2.1), which means that the contact temperature in the former case is much higher than in the latter case (2.27). This is the reason why tiles are perceived as 'colder' compared to timber parquet, despite the fact they both have the same temperature.

2.2.6 Half-space material with the harmonic change of temperature

In the second theoretical model, we assume that the half-space ($z \geq 0$) is occupied by a material of known thermal conductivity λ, density ρ and specific heat capacity c. We also assume that the boundary condition is a harmonic variation of the temperature at the surface, $z = 0$, as shown in Fig. 2.10,

$$\theta(0, t) = \bar{\theta} + \Delta\theta \sin\left(\frac{2\pi}{T}t\right),$$

where $\bar{\theta}$ is the average temperature, $\Delta\theta$ is the temperature amplitude and T is the period of temperature change (time required for the temperature to make one cycle).

Table 2.1: Diffusive properties of most common building-related materials. The period of temperature variation is taken to be 24 h for building materials and 365.24 d for soils.

Material	$a / 10^{-6} \frac{\mathrm{m}^2}{\mathrm{s}}$	$b / \frac{\mathrm{W}\,\mathrm{s}^{\frac{1}{2}}}{\mathrm{m}^2\,\mathrm{K}}$	δ/cm	$\frac{t_0}{x} / \frac{\mathrm{h}}{\mathrm{cm}}$
		$T = 24\,\mathrm{h}$		
expanded polystyrene	0.80	39	14.9	0.26
mineral wool	0.34	60	9.7	0.40
timber	0.16	320	6.7	0.57
brick, solid †	0.44	1200	11.1	0.35
brick, perforated †	0.32	390	9.4	0.40
gypsum plasterboard	0.30	380	9.1	0.42
concrete, medium density	0.75	1910	14.4	0.27
soda lime glass	0.53	1370	12.1	0.32
ceramic, porcelain	0.67	1580	13.6	0.28
steel	14.25	13 250	62.6	0.06
stone, crystalline	1.25	3130	18.5	0.21
stone, sedimentary	0.88	2450	15.6	0.24

Material	$a / 10^{-6} \frac{\mathrm{m}^2}{\mathrm{s}}$	$b / \frac{\mathrm{W}\,\mathrm{s}^{\frac{1}{2}}}{\mathrm{m}^2\,\mathrm{K}}$	δ/m	$\frac{t_0}{x} / \frac{\mathrm{d}}{\mathrm{m}}$
		$T = 365.24\,\mathrm{d}$		
soil, sand/gravel	0.98	2020	3.1	17
soil, clay/silt	0.48	2170	2.2	24

The solution of equation (2.26) is

$$\theta(z,t) = \bar{\theta} + \Delta\theta \, e^{-z/\delta} \sin\left(\frac{2\pi}{T}(t - t_0)\right), \qquad (2.29)$$

where

$$\delta = \sqrt{\frac{Ta}{\pi}}$$

is the *periodic penetration depth*, and

$$t_0 = \sqrt{\frac{T}{4\pi a}}\, z$$

is the *time delay*. The validity of the solution can be checked by inserting (2.29) into (2.26).

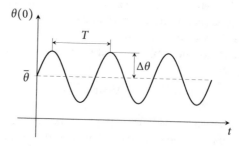

Figure 2.10: Boundary condition: harmonic variation of the temperature at the surface of the material.

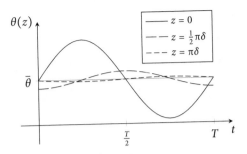

Figure 2.11: Temporal variation of the temperature for the half-space material with a harmonic change of temperature at the surface. Three depths, corresponding to the three dots shown in Fig. 2.12, are presented.

The temporal component of the solution is displayed in Fig. 2.11. With increasing depth, the temperature oscillation amplitude decreases, and the time delay increases.

The spatial component of the solution is displayed in Fig. 2.12. On the left side, we can see how the temperature oscillation 'travels' into the interior of the material with decreasing amplitude and increasing time delay. Note

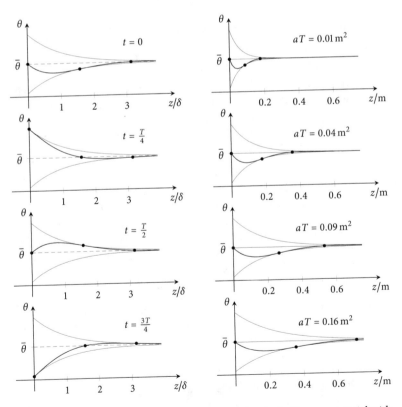

Figure 2.12: Spatial variation of the temperature for the half-space material with a harmonic change of temperature. On the left side, the temperature time evolution is shown. On the right side, the aT dependence of the spatial distribution at the initial moment is displayed.

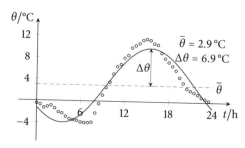

Figure 2.13: Actual temperature profile for a cloudy winter day in continental climate (open dots) and approximated harmonic function.

that the second dot is always delayed for $T/4$ against the first dot, and the third dot for $T/2$ against the first dot. This means that for a material that is deep enough, there exists a depth at which the temperature will be lowest for the highest temperature on the surface, and vice versa.

On the right side of Fig. 2.12, we can observe how smaller thermal diffusivity (or period) corresponds to a stronger reduction of amplitude and a larger time delay at the same depth z. This confirms the conclusion in Section 2.2.5 that the material with the *smaller thermal diffusivity* resists the temperature change *within* more strongly.

There are two practical applications of this theoretical model:

1. *Heat transfer dynamics within the building component.* The most important temperature cycle of interest is the daily temperature cycle, whose period is $T = 24$ h (Fig. 2.13). The building component absorbs part of the entering heat during the day when the outside temperature is high, and then releases that heat during the night when the outside temperature is low. This damps the effect of the external temperature oscillation in the internal environment.

> **Info box**
>
> Materials with small thermal diffusivity damp spatial and temporal temperature variations more intensely.

The largest effect is obtained for small thermal diffusivity, that is, for small thermal conductivity λ and large volumetric heat capacity ρc (2.24). Namely, as pointed out before, materials with small thermal conductivity hinder heat transfer, whereas materials with large volumetric heat capacity damp temperature change. However, materials possess either high thermal conductivity and volumetric heat capacity or small thermal conductivity and volumetric heat capacity, so thermal diffusivity variation is rather small (Table 2.1).

The thermal diffusivity is smallest for timber, so timber is a perfect material for building components that are composed of a single material. This is demonstrated in Table 2.1 where periodic penetration depths and time delays for the most common building materials are presented. Usually, the effect of 'small' thermal diffusivity for nontimber building components is achieved by combining two different materials, for example, an insulating part that provides small thermal conductivity and a massive part (concrete, brick) that provides large volumetric heat capacity.

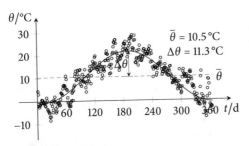

Figure 2.14: The actual temperature profile for a year in a continental climate (open dots) and the approximated harmonic function.

2. *Temperature of the ground.* The most important temperature cycle of interest is the yearly temperature cycle, whose period is T = 365.24 days (Fig. 2.14). This situation is important for studying heat losses in the building due to contact with the ground.

Furthermore, (2.29) and Fig. 2.12 indicate that for large depths, the temperature is equal to the average yearly temperature. If the average yearly temperature exceeds $0\,°C$, there is a certain depth below which the ground never freezes and if the average yearly temperature is below $0\,°C$, there is a certain depth below which the ground never melts—permafrost.

It should also be noted that in our considerations above we have assumed that the air temperature θ_e is equal to the temperature at the surface of the building element or the ground θ_{se}. As we will see in Chapter 3.1.1 on page 71, this is not the case, but the two temperatures can be related using Newton's law of cooling (2.31) and radiative heat exchange formula (2.49)

$$q = h_r(\theta_{se} - \theta_e) + h_c(\theta_{se} - \theta_e) = h(\theta_{se} - \theta_e),$$

where $h = 1/R_{se}$ is the convective-radiative surface coefficient. It can be shown that we can still use the same solution (2.29), albeit with different amplitude and time delay

$$\theta(z,t) = \bar{\theta} + A'\Delta\theta\, e^{-z/\delta} \sin\left(\frac{2\pi}{T}(t - t_0 - t_0')\right),$$

where the additional amplitude factor A' is

$$A' = \frac{1}{\sqrt{\left(1 + \frac{b}{h}\sqrt{\frac{\pi}{T}}\right)^2 + \left(\frac{b}{h}\sqrt{\frac{\pi}{T}}\right)^2}}$$

and the additional time delay t_0' is

$$t_0' = \frac{T}{2\pi} \arctan\left(\frac{\frac{b}{h}\sqrt{\frac{\pi}{T}}}{1 + \frac{b}{h}\sqrt{\frac{\pi}{T}}}\right).$$

Larger thermal effusivity results in a smaller additional amplitude factor and a larger additional time delay. This confirms the conclusion in Section 2.2.5 that the material with the *larger thermal effusivity* resists the temperature change *at the surface* more strongly. The effect can be significant in the case of the building element. For example, for concrete we get $A' = 0.65$ and $t_0' = 1.2\,\text{h}$.

This model has limited value for real building components because it does not take into account that they are not infinitely deep, they consist of several different layers and that on the other side, there are also other sources of heat that influence dynamic heat transfer. The standard ISO 13786 [43] prescribes a calculation method for the dynamic heat transfer between two thermal zones separated by a multilayer building element, provided that both thermal zones have a harmonic temperature variance with the same period. One of the zones can be the external environment, which makes this method useful for studying the daily heat losses of the building. For all other real problems it is necessary to solve the heat diffusion equation (2.26) numerically.

2.3 Convection

Heat *convection* is a mechanism of heat transfer facilitated by microscopic particles, including bulk movement of particles (advection). This mechanism appears only in fluids (see Section 1.1 on page 1.1), that is, liquids and gases, because in order to have bulk movement of particles, they must be able to move freely. Besides *particle advection*, heat transfer is also caused by *collisions between particles* and *particle diffusion*; therefore, in principle, convection includes conduction.

> **Info box**
>
> Convection is facilitated by travelling particles; therefore, it occurs in liquid and gaseous states.

Convection can be divided into two categories:

1. *Natural convection* appears when the bulk movement of particles is facilitated by some natural process, for example, buoyancy.

2. *Forced convection* appears when the bulk movement of particles is facilitated by a device, for example, a fan for gases or a pump for liquids.

A practical example of convection is the central heating system in a building (Fig. 2.15). The circulation of water in the pipes system corresponds to forced convection. Water is heated in the boiler and then forced by pump to the radiators, where it is cooled down and than returned to the boiler. The circulation of the air in the room corresponds to natural convection. First, air is warmed near the radiator. Because warmer air has lower density, it rises due to buoyancy. Moving through the room, the air eventually cools down, drops to the floor and returns back to the radiator. The circulation of air is entirely due to natural processes—no device is required.

An illustration of the importance of the convection regards thermal insulators. The primary ingredient of the most common insulating materials, expanded polystyrene (EPS), extruded polystyrene (XPS), stone and glass wool, is air (Fig. 2.2 on page 33). This is not surprising because still air is a good thermal insulator (see Table A.3 on page 277). However, as shown in the case of air in the room, circulating air can transfer considerably larger quantities of heat, so it is essential to prevent its movement.

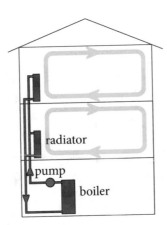

Figure 2.15: Central heating system. Forced convection is represented by the movement of water (strong colours) and natural convection by movement of air (faint colours).

This is done either by closing the air off within small pockets of polystyrene bubbles (EPS and XPS) or by hindering the movement of air by dense fibres (stone and glass wool). Note that due to the polystyrene/fibre presence, the thermal conductivity of insulating materials is somewhat larger than the thermal conductivity of the air itself.

2.3.1 Newton's law of cooling

When a radiator is opened in a cold room, air starts circulating between the warm radiator and the cold objects in the room, as shown in Fig. 2.15. This corresponds to dynamic heat transfer because the temperature of the objects and the air in the room is increasing. Eventually, the heat entering the room by the radiator is equal to the heat leaving the room through external building elements. This corresponds to steady heat transfer as the temperature of the objects in the room becomes time independent. In this case we can assume that all objects in the room, including the air, except the radiator and external building elements, have the same temperature. Therefore, the temperature gradients, that is, the temperature spatial variations, appear only in the thin air layer next to surfaces of the radiator and next to the external building elements.

Generally, in the steady heat transfer situation, the temperature gradients appear only in the thin layer of the fluid next to the solid surface. Because heat transfer is possible only in the presence of temperature gradients, *convection is then the heat transfer between the solid surface and the fluid bulk.*

> **Info box**
>
> In the steady heat transfer situation, convection is the heat transfer between the solid surface and the fluid bulk.

We will take a closer look at the microscopic picture of the convection. For simplicity, however, we will consider only the case in which the temperature of the surface is higher than the temperature of the fluid, as shown in Fig. 2.16.

The left side of Fig. 2.16 presents the microscopic picture of the forced convection. The fluid bulk velocity is v_0, and the fluid bulk temperature is θ_0.

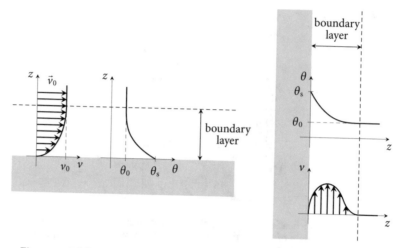

Figure 2.16: Microscopic picture of forced convection (left) and natural convection (right). In the boundary layer, the temperature and velocity of particles are gradually changing.

On the other hand, the solid surface temperature, as well as the fluid temperature next to the surface, is $\theta_s > \theta_0$. Due to the friction, the velocity of the fluid next to the surface is equal to zero. Furthermore, because of the viscosity (friction between particles of fluid), velocity of fluid gradually changes from 0 to v_0, creating a boundary layer. In an approximately equally thick boundary layer, the temperature gradually changes from θ_s to θ_0.

The right side of Fig. 2.16 presents the microscopic picture of the natural convection. Because there is no external coercion, the fluid bulk velocity is zero, and the fluid bulk temperature is θ_0. On the other hand, the solid surface temperature, as well as the fluid temperature next to the surface, is $\theta_s > \theta_0$. Due to the higher temperature, the fluid next to the surface has lower density, and buoyancy forces it up. Nevertheless, due to the friction, the velocity of fluid next to the surface is also equal to zero. Due to the viscosity (friction between particles of fluid), the velocity of the fluid gradually changes from zero to the maximum value and back to zero, creating a boundary layer. In an approximately equally thick boundary layer, the temperature gradually changes from θ_s to θ_0.

In the case of central heating, forced convection corresponds to the heat transfer from the pipe to the water in the boiler (surface is warmer than fluid) and from the water to the radiator (surface is colder than fluid). On the other hand, natural convection corresponds to the heat transfer from the radiator to the air (surface is warmer than fluid) and from the air to the external wall during the winter (surface is colder than fluid).

It is also possible to have a combination of forced and natural convection. When water is forced through a vertical pipe, forced convection due to the water pump is combined with natural convection due to buoyancy. On the other hand, when a fan is put near the radiator in the room, natural convection due to buoyancy is combined with forced convection due to the

Table 2.2: Typical values of the convective surface coefficient depending on the fluid type [1].

$h_c \big/ \frac{W}{m^2\,K}$	Type of Convection	
	Natural	Forced
gas	1–25	25–250
liquid	50–1000	100–20 000

Table 2.3: Convective surface coefficients in civil engineering [27].

$h_c \big/ \frac{W}{m^2\,K}$	Direction of Heat Flow		
	Upward	Horizontal	Downward
internal, h_{ci}	5.0	2.5	0.7
external, h_{ce}	$4+4v$	$4+4v$	$4+4v$

fan. In either of these cases, the mechanism of heat transfer is extremely complex and beyond the scope of this book.

Regardless of the convection type, the heat transfer rate is proportional to the area of the surface A and the temperature difference between the surface and fluid bulk $\theta_s - \theta_0$. This statement is called *Newton's law of cooling* and can be written as

$$\Phi = A\,h_c(\theta_s - \theta_0), \tag{2.30}$$

where constant h_c $(W/(m^2\,K))$ is called the *convective surface coefficient*. This statement also can be written in terms of the density of heat flow rate (2.3):

$$q = h_c(\theta_s - \theta_0). \tag{2.31}$$

> **Info box**
>
> Convective heat flow is proportional to the temperature difference between the solid surface and the fluid bulk.

Typical values of the convective surface coefficient are listed in Table 2.2. Note that the coefficient for gases is considerably smaller than the one for liquids, and the same applies to natural convection as compared to forced convection.

The convective surface coefficient for civil engineering cases are specified in ISO 6946 [27]. Its values for internal surfaces, or external surfaces adjacent to a well-ventilated air layer, are shown in Table 2.3. At external surfaces, the coefficient can be calculated using the expression

$$h_{ce} = 4 + 4v,$$

where v (m/s) is the wind speed adjacent to the surface.

Note that in the winter, perception of coldness is increased with wind speed. This is due to the fact that wind increases convective heat flow from human skin to the air and increases the cooling of exposed body parts.

Figure 2.17: Details of a domestic radiator. In order to increase the heat flow rate, the metal-to-air contact area is increased by the flat design, introducing sections and columns, and adding convector fins.

2.3.2 Heat and enthalpy exchangers

Convection is the principal physical mechanism behind *heat exchangers*, that is, devices that transfer heat between two or more fluids. In civil engineering it is common to encounter two types of heat exchangers:

- *Radiators* transfers heat from liquid water to air through two convective processes: first from water to an internal metal surface and then from an external metal surface to the air. The most obvious way to increase its efficiency, that is, transferred heat, is by increasing the contact area (2.30). This is especially important for metal-to-air heat transfer because the metal-to-air convective surface coefficient is much smaller than the metal-to-water coefficient (Table 2.2). The contact area is increased by a large flat design, by introducing sections and columns, and by adding convector fins, that is, the zigzagging metal strips welded to the pipes that transport liquid water (Fig. 2.17). Heat transfer can be further increased by changing from natural to forced air convection, so fans are often used to bolster the efficiency of building convector heaters and vehicle radiators.

- *Recuperators* recover energy by transferring heat between the exhaust air leaving the building and the fresh air coming into it (Fig. 2.18). This way the temperature of the fresh air is brought closer to the internal temperature and less energy is being lost. We will elaborate on recuperator application in Section 3.4 on page 111.

The efficiency of heat exchangers can be easily evaluated when two identical fluids of the same mass and the same specific heat capacity are considered. In the ideal case, the temperature of the first, initially colder fluid increases from its own inlet temperature θ_{i1} to the inlet temperature of the second fluid θ_{i2} and vice versa (Fig. 2.18). The ideal heat transfer is therefore

$$Q_{id} = mc(\theta_{i2} - \theta_{i1}).$$

However, the outlet temperatures of the first and second fluids are θ_{o1} and θ_{o2}, respectively. The real heat transfer is therefore

$$Q_{re} = mc(\theta_{o1} - \theta_{i1}) = mc(\theta_{i2} - \theta_{o2}).$$

Figure 2.18: The simple model of the recuperator consists of two metal tubes in thermal contact. During the winter period, the initially warmer exhaust air and the initially colder fresh air flow through the tubes in opposite directions. In this process heat is transferred from the exhaust air to the fresh air. The inlet temperature of the fresh air is θ_{i1} and the outlet temperature θ_{o1}. The inlet temperature of the exhaust air is θ_{i2} and the outlet temperature θ_{o2}. Note that $\theta_{o1} < \theta_{i2}$ and $\theta_{o2} > \theta_{i1}$.

The efficiency η is the ratio of two, which gives

$$\eta = \frac{\theta_{o1} - \theta_{i1}}{\theta_{i2} - \theta_{i1}}. \tag{2.32}$$

The concept of recuperators, that is gas to gas heat exchangers, is closely related to the concept of *enthalpy exchangers*. In enthalpy exchangers, the metallic barrier between exhaust air and fresh air is replaced by the water-permeable membrane through which water vapour is transferred by diffusion (Section 4.5 on page 137). This modification reduces the efficiency of *sensible* heat transfer as expressed in (2.32). However, when water vapour is transferred from the warmer to the colder air, the total energy transfer can be increased because gaseous water molecules carry a high *latent* heat (Section 1.6 on page 13). Because the enthalpy accounts for both sensible and latent heat, these recuperator types are called enthalpy exchangers. The efficiency of enthalpy exchangers can be evaluated by using two fluids of the same mass to obtain

$$\eta = \frac{h_{o1} - h_{i1}}{h_{i2} - h_{i1}}, \tag{2.33}$$

where h_{i1} and h_{i2} are the inlet specific enthalpies of two gases and h_{o1} is the outlet specific enthalpy of the initially colder gas.

2.4 Radiation

Radiation is a mechanism of heat transfer that does not require matter and can occur even in a vacuum (absence of matter). Radiation in general implies many forms of energy transfer, but with heat transfer, it applies exclusively to *electromagnetic waves*. Because electromagnetic waves are emitted by all bodies (whose temperature is above absolute zero), they represent an important contribution to heat transfer.

A typical example of heat transfer by radiation occurs between the Sun and the Earth. Because these two objects are separated by a vacuum, no other mechanism of heat transfer can occur.

Info box

Radiation is facilitated by electromagnetic waves; therefore it primarily occurs in gases and vacuums.

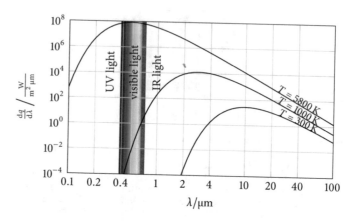

Figure 2.19: The spectral density of heat flow rate for three prominent temperatures, most notably Sun temperature and room temperature. Only objects at very high temperatures can radiate in the visual spectrum.

Bodies emit electromagnetic waves in a broad spectrum (wide range of wavelengths), and the density of heat flow rate depends on the body temperature. We will first address the radiation of *black bodies*, which are idealised physical objects that absorb all incident electromagnetic radiation. For a black body at temperature T, the emitted density of heat flow rate per wavelength λ or *spectral density of heat flow rate* $dq/d\lambda$ (W/m) is described by Planck's law

$$\frac{dq}{d\lambda}(\lambda, T) = \frac{2\pi h c_0^2}{\lambda^5 \left[\exp\left(\frac{hc_0}{\lambda k T}\right) - 1\right]},$$

where $c_0 = 2.998 \times 10^8$ m/s is the speed of light in a vacuum, $h = 6.626 \times 10^{-34}$ J s is Planck's constant and $k = 1.381 \times 10^{-23}$ J/K is Boltzmann's constant.

> **Info box**
>
> Objects at the higher temperature radiate shorter wavelengths.

The spectral density of heat flow rate for three temperatures, most notably the temperature of the Sun $T \approx 5800$ K and room temperature $T \approx 300$ K, is shown in Fig. 2.19. Note that at higher temperatures, the total emission is larger, while the spectrum moves to shorter wavelengths.

Bodies at temperatures similar to the temperature of the Sun emit radiation with a maximum in the visual spectrum, $4\,\mu m > \lambda > 7.5\,\mu m$. Incandescent light bulbs mimic the Sun in that a filament wire within the bulb is heated by electric current to temperature $T \approx 2700$ K (Fig. 2.20). At even lower temperatures—the typical centre of the fire and steel forging temperature is $T \approx 1200$ K—only the red part of the visual spectrum is emitted. Ordinary objects at room temperatures primarily emit infrared radiation and hardly any visible light, which is why we cannot see them without the presence of light sources.

> **Info box**
>
> Objects at very high temperatures are visible because they produce the light, while other objects are visible because they reflect the light.

Usually, the most important physical quantity of interest is the total density of heat flow rate. We can get its value for the black body by integrating

Figure 2.20: An incandescent light bulb emits visual, short wavelength radiation due to the high temperature of the filament, which is heated by an electric current.

Planck's law for all wavelengths

$$q(T) = \int\limits_0^\infty \frac{\mathrm{d}q}{\mathrm{d}\lambda}(\lambda, T)\mathrm{d}\lambda$$

$$\implies q(T) = \sigma\, T^4. \tag{2.34}$$

The obtained expression is called the *Stefan-Boltzmann law*, and constant $\sigma = 5.670 \times 10^{-8}\ \mathrm{W/(m^2\,K^4)}$ is called the Stefan-Boltzmann constant.

> **Info box**
>
> Objects at higher temperatures radiate more intensely.

Example 2.1: Solar constant.

Calculate the total heat flow rate of the Sun, assuming that the temperature of Sun's surface is $T_{\mathrm{Sun}} = 5780$ K and Sun's radius is $r_{\mathrm{Sun}} = 6.96 \times 10^5$ km. Calculate the value of the *solar constant*—density of heat flow rate at the outer edge of the Earth's atmosphere—if the distance from the Sun is about $r = 1.50 \times 10^8$ km. Assume that the Sun is a perfect black body.

The density of heat flow rate of the Sun is (2.34)

$$q_{\mathrm{Sun}} = \sigma\, T_{\mathrm{Sun}}^4,$$

whereas the total flow rate is (2.3)

$$\Phi_{\mathrm{Sun}} = A_{\mathrm{Sun}}\, q_{\mathrm{Sun}} = 4\pi r_{\mathrm{Sun}}^2\, \sigma\, T_{\mathrm{Sun}}^4 = 3.85 \times 10^{26}\ \mathrm{W}.$$

We assume that radiation is spreading uniformly through the space, so the heat flow rate through any element of an imaginary spherical surface centred at the Sun must be the same. To get the heat flow rate on distance r from the Sun, we must therefore divide the total heat flow of the Sun with the area of the spherical surface of the same radius (2.3):

$$q = \frac{\Phi_{\mathrm{Sun}}}{A} = \frac{\Phi_{\mathrm{Sun}}}{4\pi r^2} = 1360\ \frac{\mathrm{W}}{\mathrm{m}^2}.$$

Note that the heat flow rate is considerably smaller at the surface of the Earth because the atmosphere absorbs and reflects a part of the incident radiation.

The whole Earth is in principle irradiated by the same density of heat flow rate; however, some parts of the Earth are considerably colder than others. The warmest parts are those where the Sun is close to the zenith, that is, for the angle of incidence 0°. For parts of the Earth north and south of this position, the angle of incidence increases, so the same area of the Earth's surface gets a smaller heat flow rate (2.15). Further, the width of the atmosphere that has to be penetrated is larger there. This principle also explains the seasonal variation in temperatures.

Radiation also explains why mornings after clear sky nights are colder than mornings after cloudy nights. During the night, the Earth's surface radiation is much more intensive than the incoming radiation from stars. When radiation is not obstructed by the clouds, lost energy is much larger.

2.4.1 Grey bodies

So far we have been concerned with the idealised black body. In reality, most objects are *grey bodies* because part of the incident radiation is absorbed, part is transmitted and part is reflected (Fig. 2.21). If we denote the heat flow rates for incident radiation Φ, for reflected radiation Φ_ρ, for absorbed radiation Φ_α and for translated radiation Φ_τ, we can define *reflectance ρ, absorptance α* and *transmittance τ* as

$$\rho(\lambda) = \frac{\Phi_\rho(\lambda)}{\Phi(\lambda)}, \tag{2.35}$$

$$\alpha(\lambda) = \frac{\Phi_\alpha(\lambda)}{\Phi(\lambda)}, \tag{2.36}$$

$$\tau(\lambda) = \frac{\Phi_\tau(\lambda)}{\Phi(\lambda)}. \tag{2.37}$$

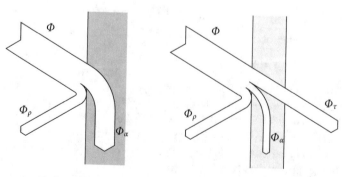

Figure 2.21: Radiation processes for nontransparent (left) and transparent (right) objects. Part of the incident radiation (Φ) is reflected (Φ_ρ), part is absorbed (Φ_α) and part is transmitted (Φ_τ).

Figure 2.22: The same scene recorded by an optical camera (left) and an infrared camera (right). Because the glass transmits short wavelength (visual) radiation but does not transmit long wavelength (infrared) radiation, the infrared emitting objects are screened by the glass. Infrared thermography will be elaborated on in Section 2.4.4.

Here we took into account that all three physical quantities depend on the radiation wavelength. Note that their value range is $0 \le \rho(\lambda), \alpha(\lambda), \tau(\lambda) \le 1$.

Because energy is conserved in the process, the sum of the heat flow rates for the reflected, absorbed and translated radiation should be equal to the density of heat flow rate for incident radiation

$$\Phi_\rho(\lambda) + \Phi_\alpha(\lambda) + \Phi_\tau(\lambda) = \Phi(\lambda).$$

Together with the definition of reflectance, absorptance and transmittance, this leads to

$$\rho(\lambda) + \alpha(\lambda) + \tau(\lambda) = 1. \tag{2.38}$$

For an ideal black body, $\alpha(\lambda) = 1$, $\rho(\lambda) = \tau(\lambda) = 0$ for all wavelengths.

A common material with unique and prominent radiation properties is glass. Its transmittance for short wavelength (visual) radiation is rather large, $\tau \approx 0.85$, whereas transmittance for long wavelength (infrared) radiation is small, $\tau \approx 0$. We can demonstrate these features by photographing scenes behind glass using optical and infrared cameras, as shown in Fig. 2.22. For the infrared camera, infrared-emitting objects are screened by the glass.

The radiant properties of glass (Fig. 2.23) also explain the greenhouse effect. The Sun radiates mostly within the short wavelength spectrum, $\lambda < 3\,\mu m$, and this radiation is transmitted through the window glass from the exterior to the interior of the building (Fig. 2.24). The Sun's radiation is absorbed by objects in the building, increasing their temperature. Because the objects are at a much lower temperature, they radiate mostly within the long wavelength spectrum, $\lambda > 3\,\mu m$, which is not transmitted but instead absorbed by the window glass. The absorbed radiation is then partially emitted back to the interior and partly to the exterior of the

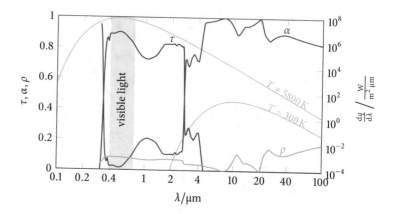

Figure 2.23: Spectral transmittance, absorptance and reflectance for common window glass [62]. Most short wavelength electromagnetic radiation (including visible light) emitted by the Sun is transmitted. Most long wavelength electromagnetic radiation is absorbed.

building. Therefore, by means of radiation, heat can enter essentially unobstructed, but it leaves the interior of the building only indirectly and partially (Fig. 2.25).

A similar effect can be observed in the Earth's atmosphere, where greenhouse gases (H_2O, CO_2, CH_4) act as a window glass.

Finally, an important property of grey bodies is that they emit less radiation than black bodies at the same temperature. To account for that phenomenon, we also define *emittance* ε as the ratio of heat flow rates, one emitted by grey body Φ to one emitted by black body Φ_0 at the same temperature T:

$$\varepsilon(\lambda) = \frac{\Phi(\lambda, T)}{\Phi_0(\lambda, T)}. \tag{2.39}$$

Figure 2.24: Quintessential greenhouse with walls of glass [63]. Temperatures within the greenhouse are elevated due to the greenhouse effect of the glass.

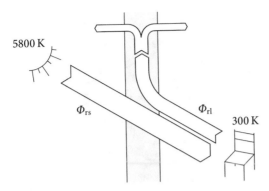

Figure 2.25: Greenhouse effect. Short wavelength radiation emitted by the Sun can enter essentially unobstructed, whereas long wavelength radiation emitted by room objects can escape the interior of the building only indirectly and partially.

2.4.2 Net radiation exchange between surfaces

We have thus far only discussed radiation emitted by a single body. Now we will pay more attention to the problem of radiative exchanges between two or more bodies.

In conduction and convection, heat is transferred only from higher to lower temperatures. However, because all bodies radiate, radiative heat is always transferred in both directions—from the body with the higher temperature to the body with the lower temperature and vice versa (Fig. 2.26). Our concern is the *net radiation* Φ_{net} as the difference of radiation that goes from the i-th body to the j-th body and the radiation that goes from the j-th body to the i-th body:

> **Info box**
>
> For radiative heat exchange between two bodies, energy travels both ways, so we are interested in net value.

$$\Phi_{\text{net}} = \Phi_{ij} - \Phi_{ji}.$$

Net radiation is always directed from the body with the higher temperature to the body with the lower temperature. As we pointed out in Section 1.2 on page 4, if both bodies have the same temperature and are in thermal equilibrium, the heat flow rates in both directions are equal, and the net radiative heat flow rate is zero.

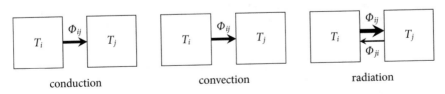

Figure 2.26: Differences between conduction, convection and radiation among two bodies, where $T_i > T_j$. In conduction and convection, heat is transferred only from higher to lower temperatures; in radiation, heat is transferred in both directions.

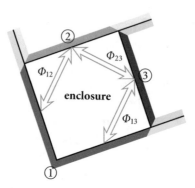

Figure 2.27: Bodies 1, 2 and 3 share a common enclosure and can therefore exchange heat by radiation. Note that only the darker shaded parts of the surfaces that confine the enclosure, not the whole surfaces of the bodies, participate in the exchange.

View factor

In order to address that problem properly, we will study the exchange between body surfaces instead of the exchange between bodies. The reason for this is quite intuitive. Energy is transferred by radiation only in an empty space, for example, in a vacuum and in gases, which means only bodies that share a well-defined enclosure can exchange heat by this mechanism. However, only those parts of surfaces that confine the enclosure, not the whole surface of the bodies, participate in the exchange, as shown in Fig. 2.27.

Net radiation exchange depends on temperatures and radiative properties of surfaces. However, it is also important to take into account that only part of the radiation that one surface emits is intercepted by another surface, which means that the net radiation exchange also depends on the surface geometries and orientations. The physical quantity that takes into consideration surface geometries and orientations is called *view factor F*. It is defined as the ratio of the heat flow rate transferred from surface i to surface j Φ_{ij} to the total heat flow rate that is emitted from surface i Φ_i, that is,

$$\Phi_{ij} = F_{ij}\Phi_i. \tag{2.40}$$

Obviously, $1 \geq F_{ij} \geq 0$. Because radiation propagates in straight lines, no radiation that leaves a convex surface can strike back (see Example 2.2). Hence, for convex surfaces, $F_{ii} = 0$. On the other hand, for concave surfaces, $F_{ii} > 0$.

All radiation emitted by surface i must be received by other surfaces that confine the same enclose, hence

$$\Phi_i = \sum_j \Phi_{ij} = \sum_j F_{ij}\Phi_i.$$

Cancelling out Φ_i on both sides, we obtain the *summation rule* stating

$$\sum_j F_{ij} = 1. \tag{2.41}$$

The net radiation exchange between two black surfaces at the same temperature should be zero:

$$\Phi_{\text{net}} = \Phi_{ij} - \Phi_{ji} = F_{ij}\Phi_i - F_{ji}\Phi_j = F_{ij}A_iq_i - F_{ji}A_jq_j = 0.$$

It follows from the Stefan-Boltzmann law (2.34) that in case of the same temperature, $T_i = T_j$, the densities of heat flow rates should also be the same, $q_i = q_j$, leading to the *reciprocity rule* of

$$F_{ij}A_i = F_{ji}A_j. \tag{2.42}$$

The net radiation exchange between two black surfaces with different temperatures is therefore

$$\Phi_{\text{net}} = F_{ij}A_i\sigma(T_i^4 - T_j^4). \tag{2.43}$$

Example 2.2: Radiation between two spherical surfaces.

Calculate the view factors for a two-surface enclosure limited by two concentric spherical surfaces of areas A_1 and A_2.

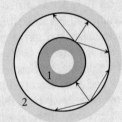

Because radiation always follows straight lines, all radiation leaving the internal surface 1 must strike the external surface 2, so

$$F_{11} = 0, \quad F_{12} = 1,$$
$$F_{11} + F_{12} = 1,$$

where the latter equation is simply confirmation of (2.41).
On the other hand, part of the radiation leaving the external surface 2 comes back to surface 2, hence

$$F_{22} > 0, \quad F_{21} < 1.$$

Using (2.42), we get

$$F_{12}A_1 = F_{21}A_2 \implies F_{21} = \frac{A_1}{A_2}.$$

Finally, from (2.41), we get

$$F_{21} + F_{22} = 1 \implies F_{22} = 1 - \frac{A_1}{A_2}.$$

We will derive the exact expression for calculating view factors in Section 8.3.4 on page 256.

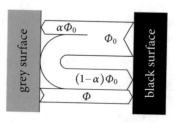

Figure 2.28: Transfer between two convex surfaces, one black and one grey, surrounding the same enclosure. The black surface emits heat flow
rate Φ_0, of which $\alpha\Phi_0$ is absorbed by the grey surface, and $(1-\alpha)\Phi_0$
is reflected back and re-absorbed by the black surface. The grey sur-
face emits heat flow rate Φ, which is completely absorbed by the black
surface.

Kirchhoff's law

By considering a simple radiative transfer between black and grey surfaces,
we can derive the relation between emittance and absorptance. We consider an enclosure surrounded by two convex surfaces: one black surface
and one nontransparent grey ($\alpha + \rho = 1$) surface in thermal equilibrium,
that is, at the same temperature T. Because surfaces are convex, all emitted radiation from the black surface strikes the grey surface and vice versa
(Fig. 2.28), that is, $F_{12} = F_{21} = 1$ and $F_{11} = F_{22} = 0$. The black surface
emits heat flow rate Φ_0, of which $\alpha\Phi_0$ is absorbed by the grey surface, and
$(1 - \alpha)\Phi_0$ is reflected back and fully absorbed by the black surface. The
grey surface emits heat flow rate Φ, which is fully absorbed by the black
surface. Because both bodies are in thermal equilibrium, absorbed and
emitted energy should be equal

$$\Phi = \alpha\Phi_0, \qquad\qquad \text{grey surface,}$$
$$\Phi_0 = \Phi + (1 - \alpha)\Phi_0, \qquad \text{black surface,}$$
$$\implies \varepsilon = \frac{\Phi}{\Phi_0} = \alpha.$$

We see that emittance and absorptance are always equal:

$$\varepsilon(\lambda) = \alpha(\lambda). \qquad\qquad (2.44)$$

That statement is called *Kirchhoff's law*.

Exchange between two grey surfaces

Although we have already calculated the net radiation exchange between
two arbitrary black surfaces (2.43), the exchange between two grey surfaces
would be much more useful. The general solution is extremely complex
and not very instructive, so we will limit the discussion to a special case
that meets two conditions:

1. The enclosure is surrounded by only two surfaces, $F_{11} + F_{12} = F_{21} + F_{22} = 1$.

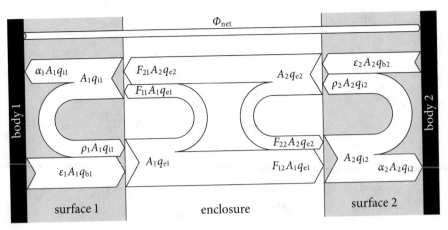

Figure 2.29: Net radiation exchange between two nontransparent grey surfaces. The individual processes and their relationships are described in the text.

2. Both surfaces are nontransparent, $\alpha_1 + \rho_1 = \alpha_2 + \rho_2 = 1$.

The schematic of the net radiation exchange between two nontransparent grey surfaces is presented in Fig. 2.29, where we assumed that the net heat transfer goes from body 1 to body 2. Note that because we have only two bodies, net heat transfer Φ_{net} is the same in the enclosure and within both surfaces. We denote q_{b1} and q_{b2} as the *black body* density of heat flow rate, q_{e1} and q_{e2} the *net emitted* density of heat flow rate and q_{i1} and q_{i2} the *net incident* density of heat flow rate for the first and second body, respectively. The net radiative exchange between the two bodies written in the enclosure is

$$\Phi_{net} = \Phi_{12} - \Phi_{21} = A_1 F_{12} q_{e1} - A_2 F_{21} q_{e2} = A_1 F_{12}(q_{e1} - q_{e2}). \qquad (2.45)$$

We want to express q_{e1} and q_{e2} in terms of body characteristics, ε_1, ε_2, q_{b1} and q_{b2}. As shown in Fig. 2.29, the emission consists of black body emitted radiation, multiplied by emittance, and the reflected part of the incident radiation:

$$q_{e1} = \varepsilon_1 q_{b1} + \rho_1 q_{i1} = \varepsilon_1 q_{b1} + (1 - \varepsilon_1) q_{i1},$$
$$q_{e2} = \varepsilon_2 q_{b2} + \rho_2 q_{i2} = \varepsilon_2 q_{b2} + (1 - \varepsilon_2) q_{i2}.$$

In order to eliminate q_{i1} and q_{i2}, we will write the net radiative exchange in both surfaces as

$$\Phi_{net} = A_1(\varepsilon_1 q_{b1} - \alpha_1 q_{i1}) = A_1 \varepsilon_1 \left(q_{b1} - \frac{q_{e1} - \varepsilon_1 q_{b1}}{1 - \varepsilon_1} \right) = \frac{\varepsilon_1}{1 - \varepsilon_1} A_1(q_{b1} - q_{e1}),$$
$$\Phi_{net} = A_2(\alpha_2 q_{i2} - \varepsilon_2 q_{b2}) = A_2 \varepsilon_2 \left(\frac{q_{e2} - \varepsilon_2 q_{b2}}{1 - \varepsilon_2} - q_{b2} \right) = \frac{\varepsilon_2}{1 - \varepsilon_2} A_2(q_{e2} - q_{b2}).$$

We can therefore rewrite the preceding expressions as

$$\Phi_{net} = \frac{q_{b1} - q_{e1}}{\frac{1 - \varepsilon_1}{A_1 \varepsilon_1}} = \frac{q_{e1} - q_{e2}}{\frac{1}{A_1 F_{12}}} = \frac{q_{e2} - q_{b2}}{\frac{1 - \varepsilon_2}{A_2 \varepsilon_2}}.$$

Using (2.13) we finally get

$$\Phi_{net} = \frac{q_{b1} - q_{b2}}{\frac{1-\varepsilon_1}{A_1\varepsilon_1} + \frac{1}{A_1 F_{12}} + \frac{1-\varepsilon_2}{A_2\varepsilon_2}},$$

$$\Phi_{net} = \frac{\sigma(T_1^4 - T_2^4)}{\frac{1-\varepsilon_1}{A_1\varepsilon_1} + \frac{1}{A_1 F_{12}} + \frac{1-\varepsilon_2}{A_2\varepsilon_2}}. \qquad (2.46)$$

2.4.3 Radiative transfer between the wall surface and the environment

Expression (2.46) can be substantially simplified for most typical civil engineering problems. The most important problem is the net radiation between the external wall of the building and its environment. For the wall external surface, the environment is the building vicinity, whereas for the wall internal surface, the environment is the rest of the room (Fig. 2.30).

We have already discussed in Section 2.3.1 that for *steady heat transfer*, we can assume that all objects in the room, including the air, except the radiator and external building elements, have the same temperature. Because the temperature of all objects in the environment is the same, the net radiation exchange for each pair of objects is zero, and the only nonzero net radiation exchange happens between the wall surface and the environment. Thus, we can consider the environment as a single surface of area A_0, average temperature T_0 and average emittance ε_0. On the other hand, the wall is a convex surface of area A_s, temperature T_s and emittance ε_s. All the radiation that leaves the surface of the wall is intercepted by the environment surface, so the view factor is $F_{s0} = 1$. Here ε_s takes into account the emittance in the most important wavelength range of the room temperature radiative spectrum, $5.5\,\mu m$–$50\,\mu m$. The single value is obtained by weighting the measured emittance with the room temperature spectral density of heat flow rate [16].

Because $A_0 \gg A_s$, from (2.46), we get

$$\Phi_{net} = A_s \sigma \varepsilon_s (T_s^4 - T_0^4).$$

Figure 2.30: Radiative transfer between the wall surface and the environment. All radiation emitted by the wall surface must be intercepted by the room environment surfaces, hence, $F_{s0} = 1$.

On the Kelvin scale, temperatures of the wall and environment surfaces do not differ significantly. We can linearise the difference in temperature to the fourth power by defining average temperature $\overline{T}$ and temperature difference ΔT, where $\Delta T \ll \overline{T}$:

$$\left.\begin{array}{l} \overline{T} = \dfrac{T_0 + T_s}{2} \\[2mm] \Delta T = T_0 - T_s \end{array}\right\} \implies \left\{\begin{array}{l} T_s = \overline{T} + \dfrac{\Delta T}{2} \\[2mm] T_0 = \overline{T} - \dfrac{\Delta T}{2}. \end{array}\right.$$

Using these expressions, we can write

$$T_s^4 = \overline{T}^4 + 4\overline{T}^3 \tfrac{\Delta T}{2} + 6\overline{T}^2 \left(\tfrac{\Delta T}{2}\right)^2 + 4\overline{T} \left(\tfrac{\Delta T}{2}\right)^3 + \left(\tfrac{\Delta T}{2}\right)^4 \approx \overline{T}^4 + 4\overline{T}^3 \tfrac{\Delta T}{2}$$

$$T_0^4 = \overline{T}^4 - 4\overline{T}^3 \tfrac{\Delta T}{2} + 6\overline{T}^2 \left(\tfrac{\Delta T}{2}\right)^2 - 4\overline{T} \left(\tfrac{\Delta T}{2}\right)^3 + \left(\tfrac{\Delta T}{2}\right)^4 \approx \overline{T}^4 - 4\overline{T}^3 \tfrac{\Delta T}{2}$$

$$\implies T_s^4 - T_0^4 = 8\overline{T}^3 \tfrac{\Delta T}{2} = 4\overline{T}^3 (T_s - T_0).$$

Taking this into account, we get

$$\Phi_{\text{net}} = A_s h_r (T_s - T_0),$$

where

$$h_r = 4\sigma\varepsilon_s \overline{T}^3 \tag{2.47}$$

is *radiative surface coefficient* h_r ($W/(m^2\,K)$) and $\overline{T}$ is the average of environment and surface temperatures. Because the temperature differences on the Kelvin and Celsius temperature scales are the same, we finally write

$$\Phi = A\,h_r (\theta_s - \theta_0). \tag{2.48}$$

> **Info box**
>
> The radiative heat exchange is proportional to the temperature difference between the wall surface and the adjacent environment.

Note the similarity of this law to Newton's law of cooling (2.30). This statement also can be written in terms of the density of heat flow rate (2.3) as

$$q = h_r (\theta_s - \theta_0). \tag{2.49}$$

The obtained expression for the radiative surface coefficient is used by standard ISO 6946 [27]. Typical values of the radiative surface coefficient in building physics are $3\,W/(m^2\,K)$ to $5\,W/(m^2\,K)$.

Example 2.3: Vacuum flask.

A vacuum flask in the shape of a cylinder consists of two polished steel vessels, the internal of height h_1 = 30.0 cm and radius r_1 = 4.0 cm and the external of height h_2 = 32.0 cm and radius r_2 = 5.0 cm. The gap between the vessels is evacuated of air, and the emittance of the vessel walls is $\varepsilon_1 = \varepsilon_2 = 0.07$. Ice of mass m = 300 g at temperature θ_i = 0 °C is inserted into vacuum flask. Calculate the ice melting time if the external temperature is θ_e = 25 °C by using an exact and linearised expression. Neglect losses through the flask neck. The specific heat of fusion for ice is q_f = 336 kJ/kg.

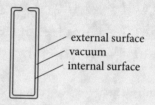

- external surface
- vacuum
- internal surface

The vacuum flask principle of operation is based on the vacuum in the gap between the two vessels. Without material particles, conduction and convection are disabled. The heat flow rate is greatly reduced because heat is transferred only by radiation. A smaller heat flow rate means that contents of the vacuum flask are cooled or warmed (depending on the contents) much slower than in other vessels.

The areas of two vessels amount to

$$A_1 = 2\pi r_1 h + 2\pi r_1^2 = 0.116 \, \text{m}^2,$$
$$A_2 = 2\pi r_2 h + 2\pi r_2^2 = 0.085 \, \text{m}^2.$$

The enclosure between the two vessels is surrounded only by two surfaces, so we can use expression (2.46). The internal surface 1 is at temperature $T_1 = T_i$ = 273 K, the external surface 2 is at temperature $T_2 = T_e$ = 298 K and because internal surface is convex, F_{12} = 1 (see Example 2.2). Taking all this into account, we get

$$\Phi = \frac{\sigma(T_i^4 - T_e^4)}{\frac{1-\varepsilon_1}{A_1\varepsilon_1} + \frac{1}{A_1} + \frac{1-\varepsilon_2}{A_2\varepsilon_2}} = -0.470 \, \text{W}.$$

The linearised form of this expression is

$$\Phi = \frac{4\sigma \overline{T}^3(T_i - T_e)}{\frac{1-\varepsilon_1}{A_1\varepsilon_1} + \frac{1}{A_1} + \frac{1-\varepsilon_2}{A_2\varepsilon_2}} = -0.469 \, \text{W}.$$

The value of the heat flow rate is negative because the internal surface receives more heat than it emits. The heat required to melt the ice is equal to (1.29, 2.2)

$$Q = m\, q_f = |\Phi|\, t \implies t = \frac{m\, q_f}{|\Phi|} = \begin{cases} 59.6 \, \text{h, (exact)} \\ 59.7 \, \text{h. (linearised)} \end{cases}$$

As pointed out before, the difference between the results for exact and linearised expressions is negligible.

2.4.4 Infrared thermography

The fact that the quantity of emitted radiation depends on the temperature of the surface can be used for noncontact temperature measurement. In civil engineering (as well as in many other fields), we are primarily concerned with temperature ranges close to room temperatures and with spatial distribution of the temperatures. This interest led to the development of a special *infrared thermography* discipline.

Objects close to room temperatures emit most of the radiation in the infrared range. Thus, thermographic instruments detect radiation in the infrared range (commonly from 8 μm to 14 μm) and produce images of that radiation.

The study of the infrared thermographic camera operation involves surfaces of three distinct entities, the object of interest at temperature T_{obj}, the instrument at temperature T_{ins} and the ambient at temperature T_{amb}. The situation is much more complicated than the case of radiative exchange between two surfaces, studied in Section 2.4.2, so we will take a simplified approach.

The object emitted density of the radiative heat flow rate is composed of thermal radiation due to the object and ambient radiation reflected from the object

$$q_{in} = \varepsilon \, \sigma \, T_{obj}^4 + (1 - \varepsilon) \, \sigma \, T_{amb}^4.$$

Here we have assumed that the object is nontransparent, $\rho = 1 - \varepsilon$, and that ambient is the perfect black body. The total heat exchange between instrument and object is

$$\Phi_{net} = \Phi_{in} - \Phi_{out} = F_{obj\text{-}ins} \, A_{obj} (q_{in} - q_{out}),$$

where $F_{obj\text{-}ins}$ is the view factor between the object and instrument, and A_{obj} is the area of the object. Assuming that the instrument is a perfect black body, the equation is transformed to

$$\Phi_{net} = F_{obj\text{-}ins} \, A_{obj} \, \sigma \left[\varepsilon \, T_{obj}^4 + (1 - \varepsilon) \, T_{amb}^4 - T_{ins}^4 \right].$$

We can calculate the view factor by taking into account the reciprocity rule (2.42) and the view factor expression (8.31)

$$A_{obj} \, F_{obj\text{-}ins} = A_{ins} \, F_{ins\text{-}obj} = \frac{\Omega_{ins\text{-}obj} \, \cos \theta_{ins}}{\pi} \, A_{ins},$$

where A_{ins} is the sensor (or aperture) area of the thermographic camera, and θ_{ins} is the angle of incidence at the instrument. This leads to

$$\Phi_{net} = \frac{\Omega_{ins\text{-}obj} \, \cos \theta_{ins}}{\pi} \, A_{ins} \, \sigma \left[\varepsilon \, T_{obj}^4 + (1 - \varepsilon) \, T_{amb}^4 - T_{ins}^4 \right].$$

Finally, the instrument signal U is proportional to the total heat exchange

$$U = C\,\Phi_{\text{net}}$$

$$= C\,\frac{\Omega_{\text{ins-obj}}\,\cos\theta_{\text{ins}}}{\pi}\,A_{\text{ins}}\,\sigma\left[\varepsilon\,T_{\text{obj}}^{4} + (1-\varepsilon)\,T_{\text{amb}}^{4} - T_{\text{ins}}^{4}\right]. \qquad (2.50)$$

Therefore, to determine its temperature, we do not have to know either the distance or the area of the measured surface, but only its solid angle and its angle of incidence, both easily obtainable from the perspective of the instrument.

Thermographic cameras measure the total heat exchange in the specified infrared range without distinguishing between different wavelengths. The obtained numerical result can be represented as a monochromatic image. However, more often results are displayed in false colour, where changes in colour rather than changes in intensity are used to represent the result. Usually, but not necessarily, the parts of the image with the highest temperatures appear white, intermediate temperatures appear in reds and yellows, and lowest temperatures appear black (Fig. 2.31). The false colours in the image are therefore completely unrelated to what we usually perceive as a colour in the visual spectrum and merely represent the temperatures.

The thermographic image is very helpful in building physics to identify the temperatures of the internal and external building surfaces. As shown in Fig. 2.31, the building on a winter day with a thicker thermal insulation has lower external surface temperatures. From equations (2.30) and (2.48), we see that lower surface temperatures and a smaller temperature differences imply smaller heat flow rate. Parts of the façade with higher temperatures reveal the weak spots in the thermal protection of the building. We will discuss this in more detail in Section 3.2.2 on page 93.

In most practical situations, the most crucial quantity in (2.50), apart from the temperature of the object, is the emittance of the surface. The exact value of this quantity is usually unknown and also differs for the surfaces captured in the same image. Fortunately, the emittance of most building components is close to one. One notable exception are all metals, which possess high reflectivity in both the visual and infrared range. In the Fig. 2.31 bottom right, the thermographic image shows that the metal roof drain pipe's apparent temperature is lower than the ambient temperature. Because of low emittance the emitted heat flow rate of the pipe is significantly smaller than the one of the surrounding items despite being at similar temperatures.

Figure 2.31: Thermographic camera image of the building on a winter day. The top images present part of the building with a thin thermal insulation, and the bottom images present part of the building with additional thermal insulation. Thicker and better implemented thermal insulation reduces the heat flow rate and lowers external surface temperatures.

Problems

2.1 A freezer of height 150 cm, width 60 cm and depth 60 cm has walls of 50 mm thick expanded polystyrene of thermal conductivity 0.040 W/(m K). The freezer maintains an internal temperature of –18 °C, and the external temperature is 30 °C. Calculate the electrical power of the freezer if the efficiency of the heat pump is 300 %. (55 W)

2.2 A cuboid vessel, which edges measure 30 cm, 35 cm and 45 cm, has walls of 20 mm thick expanded polystyrene of thermal conductivity 0.040 W/(m K). The vessel contains 2.0 kg of ice at 0 °C, and the external temperature is 30 °C. Calculate the melting time of the ice. We want to prolong melting time to 7.0 h. Calculate the thickness of the additional layer of extruded polystyrene of thermal conductivity 0.035 W/(m K). Take the specific heat of the fusion of water to be 336 kJ/kg. (3.9 h, 14 mm)

2.3 A wall has two layers. The outer layer of thermal conductivity 0.038 W/(m K) is 8.0 cm thick, and inner layer of thermal conductivity 0.16 W/(m K) is 15.0 cm thick. What is the interface temperature (temperature at the boundary of the two layers) if the temperature on the external surface of the wall is −3.0 °C and the temperature on the internal surface of the wall is 17.0 °C? Calculate the distance between the external surface of the wall and the wall interior at temperature 0.0 °C. (10.8 °C, 1.7 cm)

2.4 What temperature would gain the flat black surface on the Earth's surface, neglecting heat exchange with Earthly objects, if (a) the black surface is perpendicular to the sunbeams, (b) the angle between the black surface and sunbeams is 60°? For the Sun at zenith, the solar density of heat flow rate at the surface of the Earth is approximately 1000 W/m². What temperature would gain the flat grey surface, whose emittance is independent of wavelength? (91 °C, 78 °C; the same)

2.5 What temperature would gain the flat black surface on the Earth's surface, if the black surface is perpendicular to the sunbeams? The environment temperature is 25 °C, for the Sun at zenith the density of heat flow rate at the surface of the Earth is approximately 1000 W/m², and the average wind velocity is 2.0 m/s. What temperature would gain the flat grey surface of emittance 0.10 independent of wavelength for the same conditions? (54 °C, 29 °C)

2.6 In problem 1.13 in Chapter 1 we calculated the heat flow rate due to perspiration and respiration of an adult human being. Now calculate the net heat flow rate of radiation, and the heat flow rates of convection and conduction of a clothed inactive person in a conditioned room. For radiation and convection assume that a person is a vertical convex surface of emittance 0.97 and surface area 1.8 m², so that surface coefficients typical for building physics can be used. Take the surface temperature of a clothed person to be 28 °C and the temperature of the room to be 20 °C. For conduction assume that the surface area of the shoe sole is 5 dm² and that the thermal resistance of the shoe sole is 0.25 m² K/W. Take the temperature of the foot sole skin to be 33 °C and the temperature of the floor to be 20 °C. (83 W, 36 W, 2.6 W)

3 Heat transfer in building components

In the first part of this chapter, we will use the mechanisms that we have revealed in the previous chapter to study heat transfer in homogeneous and inhomogeneous building components. Using the obtained results, we will study heat transfer for the whole building in the second part of this chapter.

3.1 Homogeneous building components

In civil engineering, the transfer of heat combines all mechanisms: conduction, convection and radiation. Except for the physical properties of building materials, the heat flow rate depends on external and internal conditions, for example, environment temperatures. Although these conditions are generally time dependent, in this chapter, we will concentrate on *steady heat transfer*. Namely, most of the derived physical quantities (thermal resistance, thermal transmittance, heat transfer coefficient) will be independent or almost independent of external and internal conditions.

We start with *homogeneous building components*, that is, building components that are flat and whose properties change only perpendicular to the building component. This means that the problem is essentially one-dimensional—heat flows *perpendicular* to the building component surface, whereas isothermal surfaces are planar and *parallel* to the building component surface (see Section 2.2.3 on page 36).

3.1.1 Solid wall

As far as heat is concerned, the primary task of building physics is the study of heat transfer through building elements. The latter predomin-

© The Author(s), under exclusive license to Springer Nature Switzerland AG 2021
M. Pinterić, *Building Physics*,
https://doi.org/10.1007/978-3-030-67372-7_3

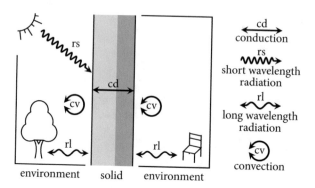

Figure 3.1: Sketch of heat transfer through a solid wall. The two environments exchange heat by conduction, convection and long wave radiation.

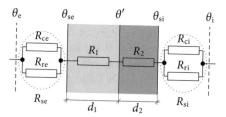

Figure 3.2: Equivalent electrical circuit for heat transfer through a solid wall.

ately consists of materials in solid state, and the only possible heat transfer mechanism in solid state is conduction. However, conduction itself cannot entirely explain the heat transfer through building elements. We can conclude that by two different ways of reasoning:

1. Temperatures on the surface of the building element and its adjacent environment (air, objects) are different. Because there is a temperature gradient in the air, additional heat transfer mechanisms must also exist. In the air, the only possible heat transfer mechanisms are convection and radiation.

2. Conduction through the solid wall transfers heat only within the solid wall, that is, between the internal surface and the external surface of the wall. If no other heat transfer mechanism existed, temperatures within the wall would eventually equalise. Therefore, additional heat transfer mechanisms must exist that keep one surface of the wall colder and the other surface of the wall warmer by transferring heat between solid wall surfaces and their adjacent environments.

Therefore, the external environment at temperature θ_e and internal environment at temperature θ_i exchange heat, and the conduction through the wall is only part of the total heat transfer process.

All heat transfer mechanisms are presented in Fig. 3.1. Note that for simplicity, we will momentarily neglect the influence of short wavelength radiation from the Sun and consider it in Section 3.1.5.

Suppose that $\theta_i > \theta_e$. Heat is transferred

- from the internal environment to the internal surface of the wall by radiation and convection,

- from the internal surface to the external surface of the wall by conduction and

- from the external surface of the wall to the external environment by radiation and convection.

Let's define *total thermal resistance* R_{tot} as the complete thermal resistance between the external and internal environments. The equivalent electrical circuit for total thermal resistance for the solid wall is shown in Fig. 3.2.

There are two mechanisms that transfer heat between the external surface of the wall and its adjacent environment: radiation with thermal resistance R_{re} and convection with thermal resistance R_{ce}. From expressions (2.31), (2.49) and (2.10), it is obvious that thermal resistances are the inverse of the corresponding surface coefficients

$$R_{ce} = \frac{1}{h_{ce}}, \; R_{re} = \frac{1}{h_{re}}.$$

Similarly, there are two mechanisms that transfer heat between the internal surface of the wall and its adjacent environment: radiation with thermal resistance R_{ri} and convection with thermal resistance R_{ci}. Thermal resistances are the inverse of the corresponding surface coefficients

$$R_{ci} = \frac{1}{h_{ci}}, \; R_{ri} = \frac{1}{h_{ri}}.$$

Because the mechanisms operate alongside, that is, parallel to each other (2.11) at the whole surface of the wall $A_{eq} = A_s$, we obtain

$$\frac{1}{R_{se}} = \frac{1}{R_{ce}} + \frac{1}{R_{re}}, \; \frac{1}{R_{si}} = \frac{1}{R_{ci}} + \frac{1}{R_{ri}},$$

$$R_{se} = \frac{1}{h_{ce} + h_{re}}, \; R_{si} = \frac{1}{h_{ci} + h_{ri}}, \quad (3.1)$$

where R_{se} (m^2 K/W) and R_{si} (m^2 K/W) are the *external surface resistance* and *internal surface resistance*, respectively. Surface resistance is thus the complete thermal resistance between the wall surface and its adjacent environment.

Info box

The thermal resistance of layers is a consequence of conductivity, whereas the thermal surface resistances are a consequence of convection and radiation.

The surface resistances depend on several conditions, such as the type of convection, emittance, velocity of wind or average temperature. ISO 6946 [27] provides not only the procedures for calculating surface coefficients (Table 2.3 on page 51 and (2.47)), but also the typical values of surface resistances, depending on the heat flow direction applicable in most practical situations. Those values are presented in Table 3.1.

Table 3.1: Wall surface resistances applicable in most practical situations [27].

$R_s \big/ \frac{m^2\,K}{W}$	Direction of Heat Flow		
	Upward	Horizontal	Downward
internal, R_{si}	0.10	0.13	0.17
external, R_{se}	0.04	0.04	0.04

Both surface resistances and thermal resistances of individual layers are in series (Fig. 3.2), so according to (2.14), the total thermal resistance R_{tot} (m^2 K/W) is

$$R_{tot} = R_{se} + R_1 + R_2 + R_{si} = R_{se} + \frac{d_1}{\lambda_1} + \frac{d_2}{\lambda_2} + R_{si}.$$

This expression can be generalised for an arbitrary number of layers in the wall:

$$R_{tot} = R_{se} + \sum_i \frac{d_i}{\lambda_i} + R_{si}. \tag{3.2}$$

In civil engineering, it is more common to use the inverse of the total thermal resistance, called the *thermal transmittance* U (W/(m^2 K)):

$$U = \frac{1}{R_{tot}} = \frac{1}{R_{se} + \sum_i \frac{d_i}{\lambda_i} + R_{si}}. \tag{3.3}$$

The heat flow rate through the building component of area A in terms of thermal transmittance is thus (2.9)

$$\Phi = AU(\theta_i - \theta_e). \tag{3.4}$$

Info box
Quantities R and k relate to a single layer, whereas quantities R_{tot} and U relate to the whole building component.

Note that thermal resistance (2.7) and coefficient of heat transfer k (2.8) relate only to the single layer, whereas the total thermal transmittance R_{tot} and thermal transmittance U relate to the whole building component.

3.1.2 Determination of characteristic temperatures

Thermal transmittance of a building component (3.3) depends only on its properties, the orientation of the building component, as well as thicknesses and thermal conductivities of constituent layers. Most importantly, it is independent of the environment temperatures, which is very convenient.

However, as will become evident in Section 4.6.1 on page 152, it is often necessary to find the characteristic temperatures of the building component (Fig. 3.2). Except for the temperature of internal environment θ_i and external environment θ_e, we are interested in wall internal surface temperature θ_{si}, wall external surface temperature θ_{se} and interface temperatures (temperatures at the boundaries of two layers) θ'. Those temperatures depend on the external θ_e and internal θ_i temperatures. Here we will introduce two methods for determining temperatures—computational and graphical.

Info box

Temperatures on the building components surface θ_{se}, θ_{si} always differ from the temperatures of adjacent environments θ_e, θ_i.

Computational method

The environment temperatures and thermal transmittance can be used to calculate the density of heat flow rate (3.4) as

$$q = \frac{\Phi}{A} = U(\theta_i - \theta_e). \qquad (3.5)$$

On the other hand, we can apply (2.6) and (2.10) for each individual layer, as well as air layers on both sides of the wall,

$$q = \frac{\lambda}{d}(\theta_h - \theta_l) = \frac{\theta_h - \theta_l}{R}, \qquad (3.6)$$

where θ_h is the higher temperature on the warmer side, and θ_l is the lower temperature on the colder side of the layer. Starting from the internal or external temperature, we can sequentially calculate all characteristic temperatures.

Example 3.1: Calculation of characteristic temperatures.

Calculate the thermal transmittance and characteristic temperatures for the vertical building component, which is composed of the following layers:

Layer	d/m	$\lambda / \frac{\mathrm{W}}{\mathrm{m\,K}}$	μ
solid layer 1 (façade plate)	0.02	1.5	50
solid layer 2 (EPS)	0.05	0.039	60
solid layer 3 (concrete)	0.15	1	120
solid layer 4 (mortar)	0.02	0.56	25

The internal temperature is $\theta_i = 20\,°\mathrm{C}$, and the external temperature is $\theta_e = -5\,°\mathrm{C}$.

Because the building component is vertical, from Table 3.1, we get $R_{si} = 0.13\,\mathrm{m^2\,K/W}$ and $R_{se} = 0.04\,\mathrm{m^2\,K/W}$. Putting that into (3.3), for thermal transmittance we get

$$U = \frac{1}{R_{se} + \frac{d_1}{\lambda_1} + \frac{d_2}{\lambda_2} + \frac{d_3}{\lambda_3} + \frac{d_4}{\lambda_4} + R_{si}} = 0.606\,\frac{\mathrm{W}}{\mathrm{m^2\,K}}.$$

We start the determination of characteristic temperatures by calculating the density of heat flow rate (3.5) as

$$q = U(\theta_i - \theta_e) = 15.1\,\frac{W}{m^2}.$$

The density of heat flow rate is constant throughout the building component. As shown in Fig. 3.3, we have to find out five characteristic temperatures. Let's start from the exterior by writing (3.6) for the external air layer as

$$q = \frac{\theta_{se} - \theta_e}{R_{se}} \implies \theta_{se} = \theta_e + R_{se}\,q = -4.39\,°C.$$

Next, we write (3.6) for solid layers as

$$q = \frac{\lambda_1}{d_1}(\theta'_1 - \theta_{se}) \implies \theta'_1 = \theta_{se} + \frac{d_1}{\lambda_1}\,q = -4.19\,°C,$$

$$q = \frac{\lambda_2}{d_2}(\theta'_2 - \theta'_1) \implies \theta'_2 = \theta'_1 + \frac{d_2}{\lambda_2}\,q = 15.22\,°C,$$

$$q = \frac{\lambda_3}{d_3}(\theta'_3 - \theta'_2) \implies \theta'_3 = \theta'_2 + \frac{d_3}{\lambda_3}\,q = 17.49\,°C,$$

$$q = \frac{\lambda_4}{d_4}(\theta_{si} - \theta'_3) \implies \theta_{si} = \theta'_3 + \frac{d_4}{\lambda_4}\,q = 18.03\,°C.$$

For a test, we can write (3.6) for the internal air layer as

$$q = \frac{\theta_i - \theta_{si}}{R_{si}} \implies \theta_i = \theta_{si} + R_{si}\,q = 20\,°C.$$

If previous calculations were precise enough, we should obtain the same temperature as it was initially specified by the problem. The calculation of the characteristic temperatures could be started from the interior as well.

The expression (3.6) can be also used to determine temperatures within layer, starting from the temperature on the edge of the layer.

Note that the expression (3.5) and multiple expressions (3.6), as written in Example 3.1, can be combined to give an explicit expression for each characteristic temperature θ'_n

$$\theta'_n = \theta_e + \left(R_{se} + \sum_{j=1}^{n} R_j\right)\frac{\theta_i - \theta_e}{R_{tot}}. \tag{3.7}$$

Using this equation, we can avoid accumulation of calculation errors.

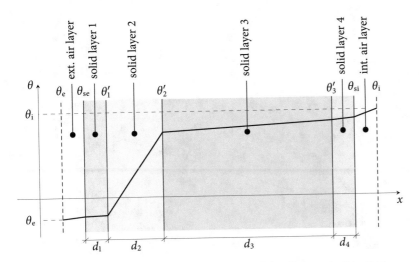

Figure 3.3: Temperature versus distance plot for a solid wall example. The thickness of the air layers is informative.

Temperature factor of the internal surface

As we will see in Section 4.5.3 on page 141, the temperature of the internal surface θ_{si} will be of special importance. From (3.7) we get

$$\theta_{si} = \theta_e + \frac{R_{tot} - R_{si}}{R_{tot}} (\theta_i - \theta_e)$$
$$= \theta_e + f_{R_{si}} (\theta_i - \theta_e).$$

This temperature is a function of time-dependant internal and external temperatures, and the *temperature independent* factor called the *temperature factor of the internal surface* $f_{R_{si}}$

$$f_{R_{si}} = \frac{\theta_{si} - \theta_e}{\theta_i - \theta_e}. \tag{3.8}$$

The temperature factor is always smaller than one, $f_{R_{si}} < 1$, and its value depends only on properties of the building component. For a homogeneous building component, its value is

$$f_{R_{si}} = \frac{R_{tot} - R_{si}}{R_{tot}}. \tag{3.9}$$

However, in more complex situations, for example, thermal bridges, the temperature factor of the internal surface has to be calculated numerically.

Note that in typical continental winter conditions, $\theta_i > \theta_e$, the larger temperature factor of the internal surface $f_{R_{si}}$ and the larger total thermal resistance R_{tot} mean a larger internal surface temperature.

> **Info box**
>
> The temperature factor of the internal surface is the temperature independent factor in the calculation of internal surface temperature.

> **Info box**
>
> The temperature of the internal surface is increased by increasing the temperature factor of internal surface and total thermal resistance, that is, by adding building insulation.

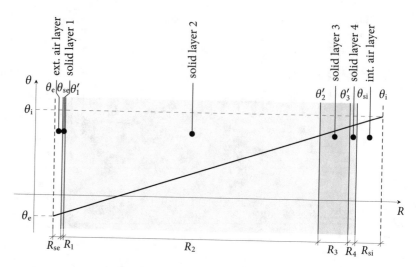

Figure 3.4: Temperature versus thermal resistance plot for the solid wall example. The function is linear, so we can use this plot to determine (read out) characteristic temperatures graphically.

Graphical method

By inspecting the temperature versus distance plot in Fig. 3.3, we can see that the temperature function has different slopes for different layers. We can explain this fact by differentiating equation (3.6) as

$$\frac{\Delta\theta}{d} = \frac{q}{\lambda} \implies \frac{\mathrm{d}\theta}{\mathrm{d}x} = \frac{q}{\lambda}.$$

The slope of the function $\theta(x)$ is described by its first derivative and therefore proportional to density of heat flow rate q (the same for all layers) and inversely proportional to thermal conductivity λ (different for each layer). It is evident that *better thermal insulators, that is, materials with smaller thermal conductivity, have steeper slopes*. Therefore, it is clear that solid layer 2 in Fig. 3.3 is a thermal insulator. We can, however, plot temperature versus thermal resistance, as shown in Fig. 3.4. In this case, from (3.6), we get

> **Info box**
>
> In the $\theta(x)$ plot, better thermal insulators (smaller thermal conductivity) have steeper slope.

$$\frac{\Delta\theta}{R} = q \implies \frac{\mathrm{d}\theta}{\mathrm{d}R} = q.$$

We see that the first derivative of $\theta(R)$ and its slope are constant; that is, the function is linear. If we plot the thermal resistances of all layers (including air layers) on abscissa, we can simply connect external temperature θ_e on one side with internal temperature θ_i on the other side. Characteristic temperatures then can be read out from the ordinate.

3.1.3 Wall with airspace

The wall with airspace problem is similar to the solid wall problem. The difference is that heat transfer through the additional layer in the wall, airspace, is facilitated by yet another combination of convection and radiation (Fig. 3.5).

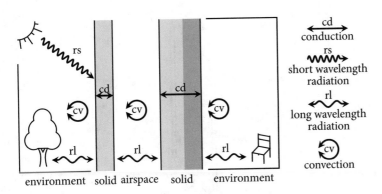

Figure 3.5: Sketch of heat transfer through a wall with airspace. Two environments exchange heat by conduction, convection and long wavelength radiation.

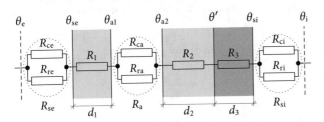

Figure 3.6: Equivalent electrical circuit for heat transfer through a wall with airspace.

Note that for simplicity we will momentarily neglect the influence of short wavelength radiation from the Sun and consider it later in Section 3.1.5.

Suppose that $\theta_i > \theta_e$. Heat is transferred

- from the internal environment to the internal surface of the wall by radiation and convection,

- from the internal surface of the wall to the internal surface of the airspace by conduction,

- from the internal surface of the airspace to the external surface of the airspace by radiation and convection,

- from the external surface of the airspace to the external surface of the wall by conduction and

- from the external surface of the wall to the external environment by radiation and convection.

Let's calculate the total thermal resistance between the external and internal environments (Fig. 3.6). Observe that this case has many similarities to the solid wall case, which was addressed in Section 3.1.1. The only significant difference is the heat transfer between opposite surfaces of the airspace. As in the air layers, the two participatory mechanisms are radiation with thermal resistance R_{ra} and convection with thermal resistance R_{ca}.

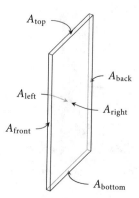

Figure 3.7: Sketch of a typical airspace. Face surfaces A_{left} and A_{right} are much larger than flanking surfaces A_{top}, A_{bottom}, A_{front}, and A_{back}. The effect of flanking surfaces can be therefore neglected and assumed that radiative heat exchange occurs only between face surfaces.

From expressions (2.30), (2.48) and (2.9), we get

$$R_{\text{ca}} = \frac{1}{h_{\text{ca}}}, \ R_{\text{ra}} = \frac{1}{h_{\text{ra}}}.$$

Because the mechanisms work alongside, that is, parallel to each other (2.11) at the whole surface of the airspace $A_{\text{eq}} = A_{\text{s}}$, we obtain

$$R_{\text{a}} = \frac{1}{h_{\text{ca}} + h_{\text{ra}}}, \tag{3.10}$$

where R_{a} (m² K/W) is the *thermal resistance of airspace*. The thermal resistance of airspace is thus the thermal resistance between the opposite surfaces of the airspace.

> **Info box**
>
> The thermal resistance of layers is a consequence of conductivity, whereas the thermal resistance of airspace is a consequence of convection and radiation.

First we will determine the radiative surface coefficient of airspace h_{ra} using the expression for net radiation exchange between two grey surfaces (2.46). Because airspace is usually large and thin, face surface areas A_{s} (Fig. 3.7) are much larger than flanking surface areas. The effect of flanking surfaces can be therefore neglected, and we can assume that all the radiation exchange occurs between convex face surfaces $F_{12} = F_{21} = 1$. Taking into account that $A_1 = A_2 = A_{\text{s}}$, we get

$$\Phi_{\text{net}} = \frac{A_{\text{s}} \sigma \left(T_2^4 - T_1^4 \right)}{\frac{1}{\varepsilon_1} + \frac{1}{\varepsilon_2} - 1},$$

where T_1 and T_2 are face surface temperatures and ε_1 and ε_2 are face surface emittances.

By linearising the difference in temperature to the fourth power, as we have already done in Section 2.4.3 on page 64, we finally get

$$\Phi_{\text{net}} = A_s \left(\frac{4\sigma \bar{T}^3}{\frac{1}{\varepsilon_1} + \frac{1}{\varepsilon_2} - 1} \right) (T_2 - T_1).$$

The part in the first bracket is obviously radiative surface coefficient

$$h_{\mathrm{ra}} = \frac{4\sigma \overline{T}^3}{\frac{1}{\varepsilon_1} + \frac{1}{\varepsilon_2} - 1}, \tag{3.11}$$

where $\overline{T}$ is the average of two face surface temperatures. Note the difference between equations (2.47) and (3.11). Whereas the former describes the radiative exchange between the surface and the environment, the latter describes the radiative exchange between the two surfaces within the airspace.

The theoretical calculation of convective surface coefficient of airspace h_{ca} is much more complicated. Standard ISO 6946 [27] specifies its calculation for unventilated airspace.

Both surface resistances, thermal resistance of airspace and thermal resistances of individual layers are in series, so according to (2.14), the total thermal resistance R_{tot} is

$$R_{\mathrm{tot}} = R_{\mathrm{se}} + R_1 + R_{\mathrm{a}} + R_2 + R_3 + R_{\mathrm{si}} = R_{\mathrm{se}} + \frac{d_1}{\lambda_1} + R_{\mathrm{a}} + \frac{d_2}{\lambda_2} + \frac{d_3}{\lambda_3} + R_{\mathrm{si}}.$$

This expression can be generalised for an arbitrary number of layers in the wall

$$R_{\mathrm{tot}} = R_{\mathrm{se}} + R_{\mathrm{a}} + \sum_i \frac{d_i}{\lambda_i} + R_{\mathrm{si}}, \tag{3.12}$$

$$U = \frac{1}{R_{\mathrm{se}} + R_{\mathrm{a}} + \sum_i \frac{d_i}{\lambda_i} + R_{\mathrm{si}}}. \tag{3.13}$$

Standard ISO 6946 [27] specifies practical engineering calculations for three types of airspaces:

1. *Unventilated air layer.* The thermal resistance of the airspace is calculated using specified expressions.

2. *Well-ventilated air layer.* The thermal resistance of the air layer and all other layers between the air layer and the external environment is neglected. Instead, the external surface resistance corresponding to still air is used.

3. *Slightly ventilated air layer.* The total thermal resistance is obtained by interpolating linearly the total thermal resistances of the unventilated and well-ventilated air layers.

3.1.4 Double-glazed window

The double-glazed window case is equivalent to the wall with airspace case. The only difference is that short wavelength radiation is transmitted in one direction directly from the Sun to the internal environment (Fig. 3.8).

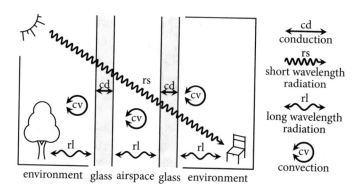

Figure 3.8: Sketch of heat transfer through a double-glazed window. Heat is transferred between the two environments by conduction, convection and long wavelength radiation. Short wavelength radiation provides one-way heat transfer between the Sun and the internal environment.

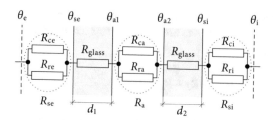

Figure 3.9: Equivalent electrical circuit for heat transfer through a double-glazed window.

Note that for simplicity, we will momentarily neglect the influence of short wavelength radiation from the Sun and consider it later in Section 3.1.5.

Suppose that $\theta_i > \theta_e$. Heat is transferred

- from the internal environment to the internal surface of the glazing by radiation and convection,

- from the internal surface of the glazing to the internal surface of the airspace by conduction,

- from the internal surface of the airspace to the external surface of the airspace by radiation and convection,

- from the external surface of the airspace to the external surface of the glazing by conduction and

- from the external surface of the glazing to the external environment by radiation and convection.

Mechanisms of heat transfer are analogous to the wall with airspace case in Section 3.1.3. The total thermal resistance R_{tot} therefore includes two glass resistances and a thermal resistance of airspace (Fig. 3.9):

$$R_{\text{tot}} = R_{\text{se}} + \frac{d_1}{\lambda_{\text{glass}}} + \frac{1}{h_{\text{ca}} + h_{\text{ra}}} + \frac{d_2}{\lambda_{\text{glass}}} + R_{\text{si}}.$$

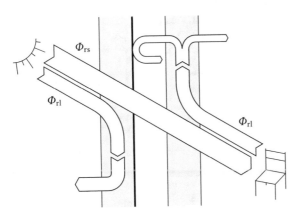

Figure 3.10: Effect of glazing pane coating. Due to the high reflectance, the coated glass pane completely reflects the outbound short wavelength radiation. The coated glass pane also does not emit any short wavelength radiation inward due to the low emittance.

The distinction regarding the wall with airspace case is that the thermal resistance of glass panes is very small. Typically, the glass is between 4 mm and 6 mm thick, whereas its thermal conductivity is among the largest in building physics (Table A.3 on page 277). For a glass thickness of 4 mm, the thermal resistance (2.7) of the glass amounts to $R_{\text{glass}} = 0.006\,\text{m}^2\,\text{K/W}$, which is far smaller than surface resistances (Table 3.1 on page 74). Therefore, the function of glass panes is primarily to separate air layers, as evident in problems 3.1 and 3.2. On the other hand, we can see that the thermal resistance of airspace will have a decisive influence on the total thermal resistance of the double glazing.

> **Info box**
>
> The thermal resistance of glass panes is negligible, so their function is only to separate air layers.

The calculation of the total thermal resistance and thermal transmittance is not as straightforward as in previous cases. Instead of calculations, ready data are provided for the most common types of airspaces and glazings in ISO 52022-2 [52].

Through tweaking of radiation and convection, the thermal resistance of airspace and consequently the total thermal resistance of double glazing can be significantly increased.

The first method of increasing thermal resistance is coating external glass panes on the internal side to obtain low emittance (low-e) glazing. The coating is very thin (in order of 10 nm), so transmittance of the short wavelength (visual) spectrum is not significantly reduced. On the other hand, reflectance in the long wavelength spectrum is significantly increased, whereas emittance and absorptance are significantly reduced. Due to the high reflectance, the coated glass pane completely reflects the outbound short wavelength radiation (Fig. 3.10). Compare that to the case in which the glass pane partially absorbs and then partially re-radiates heat towards the external environment (Fig. 2.25 on page 59). Due to low emittance, the coated glass pane also does not emit any short wavelength radiation inward.

We can quantitatively demonstrate the effect by using expression (3.11). For normal glass, emittance is equal to $\varepsilon_1 = \varepsilon_2 = 0.89$, and by taking the

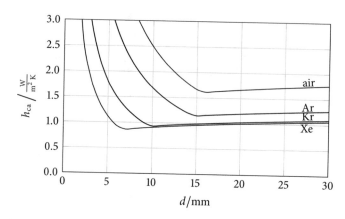

Figure 3.11: The convective surface coefficient for glazing airspace calculated according to EN 673 [11] ($\overline{T}$ = 283 K, $\Delta\theta$ = 15 °C) as a function of airspace thickness. For small thicknesses, conduction prevails, whereas for larger thicknesses, convection sets in. There is a certain airspace thickness for which the convective surface coefficient is minimal.

typical continental average winter temperature $\overline{\theta}$ = 10 °C, $\overline{T}$ = 283 K, we get h_{ra} = 4.12 W/(m² K) for the radiative surface coefficient. If the glazing pane is coated, emittance typically can be reduced to ε_1 = 0.04, which decreases radiative surface coefficient to merely h_{ra} = 0.20 W/(m² K) and creates a twentyfold reduction in radiation losses between the two glass panes.

> **Info box**
>
> The glazing thermal resistance can be improved by a low-e coating, which reduces radiative transfer, and by inert gas filling, which reduces convective transfer.

The second method of increasing thermal resistance is filling the space between the glass panes with various inert gases (argon, krypton, xenon, sulphur hexafluoride). This significantly reduces convective surface coefficient h_{ca} (Fig. 3.11). Note that there is a certain gas-dependent airspace thickness for which the convective surface coefficient is minimal.

The combined effect can be demonstrated using ISO 52022-2 [52]. Double glazing with 4 mm thick pane and 12 mm thick airspace has the following thermal transmittances:

- Normal glass and air filling: U = 2.8 W/(m² K)

- Normal glass and argon filling: U = 2.7 W/(m² K)

- Coated glass and air filling: U = 1.7 W/(m² K)

- Coated glass and argon filling: U = 1.3 W/(m² K)

The total thermal resistance also could be increased significantly by adding another window pane, thus creating a triple-glazed window (Fig. 3.24 on page 102). In this case, the total thermal resistance is equal to $R_{tot} = R_{se} + 2R_a + 3R_{glass} + R_{se}$.

A vacuum flask is another interesting case of double glazing with coated glass. Furthermore, the area between the internal and external glass of the vacuum flask is evacuated to completely stop convection, leaving radiation the sole mechanism of heat transfer.

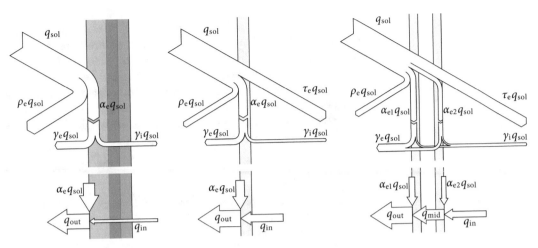

Figure 3.12: Solar gains for a nontransparent building element (left), a transparent or single glazed building element (middle) and a double glazed building element (right). As shown in the sketches above, a part of the absorbed solar radiation ($\alpha_e q_{sol}$) and all of the transmitted solar radiation ($\tau_e q_{sol}$) provide additional densities of heat flow rate towards the internal environment. As shown in the sketches below, the absorbed radiation changes the densities of the heat flow rate from the internal to the external environment.

3.1.5 Solar gain

So far, we have neglected the effect of solar radiation, whose density of heat flow rate in zenith position at the Earth's surface is approximately 1000 W/m². The actual solar density of heat flow rate q_{sol} (which includes both short- and long-wavelength radiation) varies depending on the geographic latitude, the period of year and the period of day.

In Section 2.4.1 on page 56, we have already shown that the transparent building elements transmit solar radiation to the internal environment with a density of heat flow rate of $\tau_e q_{sol}$ (Fig. 3.12, top). On the other hand, the radiation absorbed by the transparent and nontransparent building element with a density of heat flow rate of $\alpha_e q_{sol}$, increases building element temperature.

For a nontransparent building element (Fig. 3.12, left) and for a single glazed building element (Fig. 3.12, middle), the solar direct transmittance τ_e and the solar direct absorptance α_e simply take into account the transmittance and absorptance of the element in the most important wavelength range of the solar radiative spectrum, 0.3 μm–2.5 μm [10, 30]. However, in the case of a double glazed building element (Fig. 3.12, right), the solar direct transmittance and absorptance must take into account the radiative properties of both panes. In particular, the solar direct transmittance τ_e takes into account not only the radiation that is directly transmitted by both panes, but also the radiation that is transmitted by the outer pane, then (multiple times) reflected between two panes and finally transmitted by the inner pane. Similarly, the solar direct absorptance of the outer pane α_{e1} takes into account not only the radiation that is directly absorbed by the outer pane, but also for the radiation that is transmitted by the outer pane, then (multiple times) reflected between two panes and finally absorbed by the outer pane, while the solar direct absorptance of the inner pane α_{e2} takes into account not only the radiation that is transmitted by the outer pane and then directly absorbed by the inner pane, but also the radiation that is transmitted by the outer pane, then (multiple times) reflected between two panes and finally absorbed by the inner pane. All solar direct transmittances and absorptances are weighted by the solar spectral density of the heat flow rate [10, 30].

In the winter period the higher temperature of the building element (Fig. 3.12, bottom)

- increases the temperature difference and consequently the density of the heat flow rate from the building element to the external environment q_{out} and

- decreases the temperature difference and consequently the density of the heat flow rate from internal environment to the building element q_{in}.

This can be perceived as an additional density of the heat flow rate towards the external environment $\gamma_e q_{sol}$ and an additional density of the heat flow rate towards the internal environment $\gamma_i q_{sol}$, both resulting from the absorbed energy (Fig. 3.12, top). Heat transfer due to solar radiation thus has two components, and both should be taken into account when studying solar gain.

Nontransparent building element

In the case of the nontransparent building element, the density of the heat flow rate of solar gain has only one component $\gamma_i q_{sol}$ (Fig. 3.12, left). We assume that the entire absorption process takes place only on the external surface, which means that the density of the heat flow rate from the internal environment to the building element is equal to the density of the heat flow rate through the building element. We can therefore write

$$q_{in} = \frac{\theta_i - \theta_{se}}{\sum_i R_i + R_{si}}, \tag{3.14}$$

$$q_{out} = \frac{\theta_{se} - \theta_e}{R_{se}}. \tag{3.15}$$

Note that q_{in} represents the actual heat losses. By writing the relationship between three densities of heat flow rate meeting at the external surface and using (3.15), we can also express it as

$$q_{in} = q_{out} - \alpha_e q_{sol} = \frac{\theta_{se} - \theta_e - R_{se} \alpha_e q_{sol}}{R_{se}}. \tag{3.16}$$

If we put (3.14) and (3.16) together, we finally get

$$q_{in} = \frac{\theta_i - \theta_e - R_{se} \alpha_e q_{sol}}{R_{se} + \sum_i R_i + R_{si}} = U(\theta_i - \theta_e - R_{se} \alpha_e q_{sol}).$$

We can rewrite this expression in a similar form to expression (3.5)

$$q = U(\theta_i - \theta_{\text{sol-air}}),\qquad(3.17)$$

if we replace the external temperature with the equivalent, *sol-air temperature* $\theta_{\text{sol-air}}$

$$\theta_{\text{sol-air}} = \theta_e + \alpha_e\, R_{se}\, q_{\text{sol}}.\qquad(3.18)$$

The sol-air temperature can also contain additional infrared radiation due to the difference between the external air temperature θ_e and apparent sky temperature θ_{sky}

$$\Delta q_{\text{sky}} = F_{\text{sky}}\, h_{\text{re}}\,(\theta_e - \theta_{\text{sky}}),$$

where F_{sky} is the view factor between the building component and the sky, for example, 1.0 for horizontal and 0.5 for vertical building components. If we add Δq_{sky} in the same way as we added $\alpha_e\, q_{\text{sol}}$ before, we get for the sol-air temperature

$$\theta_{\text{sol-air}} = \theta_e + \alpha_e\, R_{se}\, q_{\text{sol}} - R_{se}\,\Delta q_{\text{sky}}.$$

Transparent or single glazed building element

In the case of the transparent or single glazed building element, the density of the heat flow rate due to solar gain has two components, $\tau_e q_{\text{sol}}$ and $\gamma_i q_{\text{sol}}$ (Fig. 3.12, middle). We assume that the whole absorption process takes place only on the external surface, which means that the density of the heat flow rate from the internal environment to the building element is equal to the density of the heat flow rate through the building element. We can therefore write

$$q_{\text{in}} = \frac{\theta_i - \theta_{se}}{R + R_{si}},\qquad(3.19)$$

$$q_{\text{out}} = \frac{\theta_{se} - \theta_e}{R_{se}}.\qquad(3.20)$$

Note that q_{in} represents the actual heat losses. By writing the relationship between three densities of heat flow rate meeting at the external surface and using (3.20), we can also express it as

$$q_{\text{in}} = q_{\text{out}} - \alpha_e q_{\text{sol}} = \frac{\theta_{se} - \theta_e - R_{se}\alpha_e q_{\text{sol}}}{R_{se}}.\qquad(3.21)$$

If we put (3.19) and (3.21) together, we finally get

$$q_{\text{in}} = \frac{\theta_i - \theta_e - R_{se}\alpha_e q_{\text{sol}}}{R_{se} + R + R_{si}},$$

which can be written as the sum of a density of heat flow rate from the internal to the external environment without the presence of solar gain q and an additional solar density of heat flow rate towards internal environment due to the solar gain $\gamma_i q_{sol}$

$$q_{in} = q - \gamma_i q_{sol},$$

$$q = \frac{\theta_i - \theta_e}{R_{se} + R + R_{si}} = U(\theta_i - \theta_e),$$

$$\gamma_i = \frac{R_{se}\alpha_e}{R_{se} + R + R_{si}} = U R_{se}\alpha_e. \tag{3.22}$$

We define the *solar factor* g as the ratio of the additional densities of the heat flow rate to the internal environment due to solar radiation q_{add} to the density of heat flow rate of the incident solar radiation

$$g = \frac{q_{add}}{q_{sol}} = \frac{\tau_e q_{sol} + \gamma_i q_{sol}}{q_{sol}} = \tau_e + \gamma_i$$

$$\implies g = \tau_e + U \alpha_e R_{se}. \tag{3.23}$$

Note that the first part of the solar factor corresponds to the solar radiation that is transmitted through the building component and the second part corresponds to the solar radiation that is absorbed by the building component and then released to the internal environment. The additional contribution to the density of heat flow rate towards the internal environment due to solar radiation is then

$$q_{add} = g\, q_{sol}. \tag{3.24}$$

In the United States, the quantity Solar Heat Gain Coefficient (SHGC) is used instead of the solar factor. The solar gain of glazing is an important issue for the *passive house* concept, which is discussed in Section 3.4.

Standards EN 410 and ISO 9050 [10, 30] provide for the calculation of the solar gain of glazing. The standard prescribes special calculations of τ_e, α_e (0.38 μm–2.5 μm), as well as a calculation of $h_e = 1/R_{se}$ and $h_i = 1/R_{si}$ (5.5 μm–50 μm), whereby the thermal resistance of the glazing is neglected $R = 0\,\text{m}^2\,\text{K/W}$ in (3.22). This leads to the final expression

$$\gamma_i = \frac{h_i \alpha_e}{h_e + h_i}.$$

Double glazed building element

The case of the double glazed building element is somewhat more complex because both panes absorb part of the solar radiation (Fig. 3.12, right). We assume that the entire absorption process takes place only on the external surface of the outer pane and on the internal surface of the inner pane. It can be shown (see Problem 3.2) that the temperature of each pane is practically constant, so this assumption has only a negligible effect on the result. Because energy is absorbed on two planes, we must consider three heat transfers: density of heat flow rate from the building element to the external environment q_{out}, density of heat flow rate through the building element q_{mid} and density of heat flow rate from the internal environment to the building element q_{in}. They can be expressed as

$$q_{in} = \frac{\theta_i - \theta_{si}}{R_{si}}, \tag{3.25}$$

$$q_{mid} = \frac{\theta_{si} - \theta_{se}}{R + R_a + R}, \tag{3.26}$$

$$q_{out} = \frac{\theta_{se} - \theta_e}{R_{se}}. \tag{3.27}$$

Note that q_{in} represents the actual heat losses. If we write the relationship between three densities of heat flow rate meeting at the internal surface and use (3.26), we can also express it as

$$q_{in} = q_{mid} - \alpha_{e2}q_{sol} = \frac{\theta_{si} - \theta_{se} - (R + R_a + R)\alpha_{e2}q_{sol}}{R + R_a + R}. \tag{3.28}$$

Finally, if we write the relationship between three densities of heat flow rate meeting at the external surface and use (3.27), we can also express it as

$$q_{in} = q_{out} - (\alpha_{e1} + \alpha_{e2})q_{sol} = \frac{\theta_{se} - \theta_e - R_{se}(\alpha_{e1} + \alpha_{e2})q_{sol}}{R_{se}}. \tag{3.29}$$

Using (2.13) on (3.25), (3.28) and (3.29), we finally get

$$q_{in} = \frac{\theta_i - \theta_e - R_{se}(\alpha_{e1} + \alpha_{e2})q_{sol} - (R + R_a + R)\alpha_{e2}q_{sol}}{R_{se} + R + R_a + R + R_{si}}, \tag{3.30}$$

which can be written as the sum of a density of heat flow rate from the internal to the external environment without the presence of solar gain q and an additional solar density of heat flow rate towards the internal environment due to the solar gain $\gamma_i q_{solar}$

$$q_{in} = q - \gamma_i q_{sol},$$
$$q = \frac{\theta_i - \theta_e}{R_{se} + R + R_a + R + R_{si}} = U(\theta_i - \theta_e),$$
$$\gamma_i = \frac{R_{se}(\alpha_{e1} + \alpha_{e2}) + (R + R_a + R)\alpha_{e2}}{R_{se} + R + R_a + R + R_{si}}. \tag{3.31}$$

For the solar factor of a double glazed building element we finally get

$$g = \tau_e + U[(\alpha_{e1} + \alpha_{e2})R_{se} + \alpha_{e2}(R + R_a + R)]. \tag{3.32}$$

The standards EN 410 and ISO 9050 [10, 30] prescribe a special calculation of τ_e, α_{e1}, α_{e2} (0.3 μm–2.5 μm), as well as $h_e = 1/R_{se}$ and $h_i = 1/R_{si}$ (5.5 μm–50 μm). The thermal conductance between the external and internal surfaces of the glazing $\Lambda = 1/(R + R_a + R)$ is calculated according to EN 673 [11] (5.5 μm–50 μm). This leads to the final expression

$$\gamma_i = \frac{\frac{\alpha_{e1} + \alpha_{e2}}{h_e} + \frac{\alpha_{e2}}{\Lambda}}{\frac{1}{h_e} + \frac{1}{\Lambda} + \frac{1}{h_i}}.$$

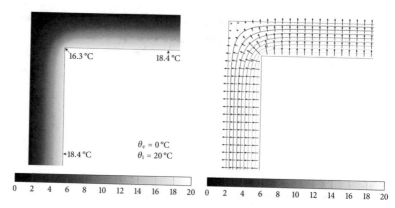

Figure 3.13: Temperatures (left), isotherms and density of heat flow rate vectors (right) for a linear geometric thermal bridge. Isotherms are not planes, and a direction of the heat flow rate changes in the vicinity of the thermal bridge. For a large distance from the thermal bridge, the situation becomes homogeneous. Results were obtained using FreeFem++ [64].

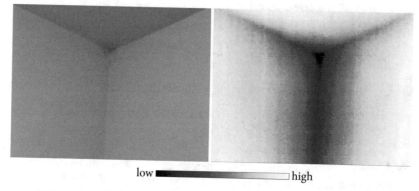

Figure 3.14: Thermographic analysis of external linear and point geometric thermal bridges from inside: temperatures in the thermal bridge vicinity are reduced, boosting mould growth.

3.2 Inhomogeneous building components and thermal bridges

Building components are not homogeneous if either the building component is not flat, or the properties change along the building component, or a combination thereof. In other words, the situation is no longer one-dimensional, heat flows in various directions and isotherms are not planar. Due to these complications, the exact heat flow rate can no longer be calculated analytically.

A simple example of the inhomogeneous building component is the building edge. A numerical calculation of temperatures shows that heat flows in various directions, and isotherms are not planar (Fig. 3.13). This can be confirmed by the thermographic analysis from inside, where the vicinity of the thermal bridges has reduced temperatures (Fig. 3.14).

3.2.1 Simplified calculation

When a building component is flat, and thermal conductivities within one layer differ moderately, a simplified calculation can be applied. This is especially convenient in cases where the changes in structure follow a certain pattern (Fig. 3.15, top), for example, in cases of prefabricated timber walls. The general idea is to split a building component into either several layers parallel to the surface (1, 2, 3, ...) or into several sections perpendicular to the surface (a, b, c, ...) and then calculate the total thermal resistance by assuming that

1. heat flows perpendicular to the building component surface, or

2. all isotherms are planar and parallel to the building component surface.

The total thermal resistance is an *average* of the total thermal resistances obtained by both methods. The method is specified by standard ISO 6946 [27] and will be demonstrated for a simple example below.

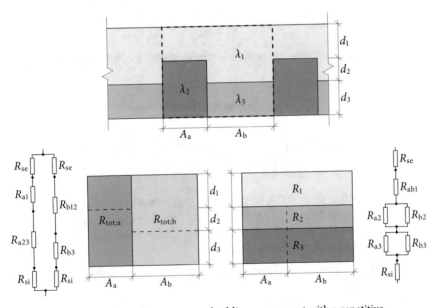

Figure 3.15: Example of an inhomogeneous building component with a repetitive pattern (top) and two simplified ways of calculating total thermal resistance (bottom). In the bottom left, we assume that heat flows perpendicular to the building component surface, and in the bottom right, we assume that isotherms are planar and parallel to the building component surface.

1. Heat flows perpendicular to the building component surface

According to this assumption (Fig. 3.15, bottom left), we first calculate the total thermal resistance of sections and then join them to get the total thermal resistance of the building component. This method provides the upper limit of the total thermal resistance.

We calculate total thermal resistance of sections $R_{\text{tot;a}}$ and $R_{\text{tot;b}}$ using (3.2):

$$R_{\text{tot;a}} = R_{se} + \frac{d_1}{\lambda_1} + \frac{d_2 + d_3}{\lambda_2} + R_{si},$$

$$R_{\text{tot;b}} = R_{se} + \frac{d_1 + d_2}{\lambda_1} + \frac{d_3}{\lambda_3} + R_{si}.$$

The upper limit of total thermal resistance is (2.11)

$$\frac{A_a + A_b}{R_{\text{tot;upper}}} = \frac{A_a}{R_{\text{tot;a}}} + \frac{A_b}{R_{\text{tot;b}}}.$$

If we define *fractional areas* of each section

$$f_a = \frac{A_a}{A_a + A_b}, \quad f_b = \frac{A_b}{A_a + A_b}, \tag{3.33}$$

we can rewrite the expression for the upper limit of total thermal resistance as

$$\frac{1}{R_{\text{tot;upper}}} = \frac{f_a}{R_{\text{tot;a}}} + \frac{f_b}{R_{\text{tot;b}}}.$$

2. All isotherms are planar and parallel to the building component surface

According to this assumption (Fig. 3.15, bottom right), we first calculate thermal resistance of layers and then join them to get the total thermal resistance of the building component. This method gives the lower limit of the total thermal resistance.

We calculate the thermal resistance of layers R_1, R_2, R_3 using (2.11) and (3.33) as

$$R_1 = R_{ab1},$$

$$\frac{A_a + A_b}{R_2} = \frac{A_a}{R_{a2}} + \frac{A_b}{R_{b2}} \implies \frac{1}{R_2} = \frac{f_a}{R_{a2}} + \frac{f_b}{R_{b2}},$$

$$\frac{A_a + A_b}{R_3} = \frac{A_a}{R_{a3}} + \frac{A_b}{R_{b3}} \implies \frac{1}{R_3} = \frac{f_a}{R_{a3}} + \frac{f_b}{R_{b3}}.$$

where R_{ab1} is thermal resistance of the first layer, R_{a2} and R_{b2} are sectional thermal resistances of the second layer and R_{a3} and R_{b3} are sectional thermal resistances of the third layer.

Because the thickness for R_1, R_{ab1} is d_1; the thickness for R_2, R_{a2}, R_{b2} is d_2; and the thickness for R_3, R_{a3}, R_{b3} is d_3, the expressions could be simplified using (2.7) as

$$\lambda_{eq;1} = \lambda_1,$$
$$\lambda_{eq;2} = f_a \lambda_2 + f_b \lambda_1,$$
$$\lambda_{eq;3} = f_a \lambda_2 + f_b \lambda_3.$$

The lower limit of the total thermal resistance is (3.2)

$$R_{tot;lower} = R_{se} + \frac{d_1}{\lambda_{eq;1}} + \frac{d_2}{\lambda_{eq;2}} + \frac{d_3}{\lambda_{eq;3}} + R_{si}.$$

Averaging total thermal resistances

In the end, we have to average the upper and lower limits of the total thermal resistances as

$$R_{tot} = \frac{R_{tot;upper} + R_{tot;lower}}{2}.$$

The upper limit of error can be estimated by

$$e = \frac{R_{tot;upper} - R_{tot;lower}}{2R_{tot}}.$$

3.2.2 Thermal bridges

A *thermal bridge* is part of the thermal envelope (within a building element or on the boundary of two different building elements), where thermal transmittance U and direction of density of heat flow rate $\vec{q}$ change significantly. Both effects can be observed in the numerical analysis of the building edge depicted in Fig. 3.13 on page 90.

Thermal bridges are usually due to the penetration of the insulation (low-conductive layer) by a highly conductive material, thus *increasing* thermal transmittance and heat losses. It is interesting to note that in buildings with thicker thermal insulation, the share of heat losses through the thermal bridges is larger because the effect of adding insulation is stronger on homogeneous than inhomogeneous building components. For this reason, special care should be taken to avoid thermal bridges or at least mitigate their contribution to the heat flow rate.

Thermal bridges (Fig. 3.16) can be divided into two main categories:

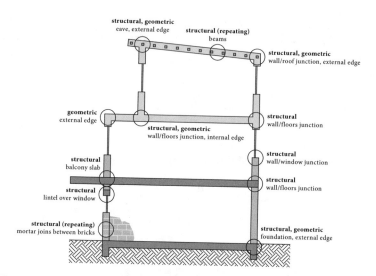

Figure 3.16: Some typical thermal bridges on the cross-section of a building. The division into geometric and structural thermal bridges is indicated.

1. *Structural thermal bridges* are parts of the thermal envelope where the uniform structure is changed by a full or partial penetration of the building envelope by materials with a different thermal conductivity and/or a change in thickness of the fabric. Usually, the thermal envelope is penetrated by another material for stability reasons, such as the foundation, balcony slab, lintel over the window or wall/floors junction, or when two different structures meet, such as the wall/window junction or wall/roof junction.

2. *Geometric thermal bridges* are parts of the thermal envelope where the difference between the internal and external areas appears, which is the result of the geometry (or shape) of the building. These bridges are placed at the junction of two planes (edge) or three planes (corner), with the planes representing walls, floors or ceilings (Fig. 3.13 on page 90).

Structural thermal bridges are often further divided into *repeating thermal bridges* and *nonrepeating thermal bridges*. Repeating thermal bridges occur following a regular pattern, such as mortar joins between bricks or construction beams. Repeating thermal bridges can be sometimes taken into account using the simplified calculation described in Section 3.2.1.

The other possible categorisation of thermal bridges takes into account their dimensionality and type of their contribution to heat flow rate:

A. *Linear thermal bridges* are those whose distribution can be simplified by a line, such as a building edge or wall/floor junction. The heat in the thermal bridge vicinity flows in two dimensions.

B. *Point thermal bridges* are those whose distribution can be simplified by a point, such as a corner or mechanical fastener. The heat in the thermal bridge vicinity flows in three dimensions.

low ▮▬▬▬▬▬▬▬▯ high

Figure 3.17: Thermographic analysis of thermal bridges: wall/floors junctions, lintels over windows, wall/window junctions and external and internal edges.

Note that both structural and geometric thermal bridges can be either linear or point.

Thermal bridges can be analysed through infrared thermography (see Section 2.4.4 on page 67), that is, by studying the temperatures of the building surfaces (Fig. 3.17). Because the heat flow rate in winter can be expressed by

$$q = \frac{\theta_i - \theta_{si}}{R_{si}} = \frac{\theta_{se} - \theta_e}{R_{se}},$$

higher external surface temperatures and lower internal surface temperatures indicate a larger heat flow rate.

> **Info box**
>
> Thermal bridges are detected by higher external surface temperature and lower internal surface temperatures (winter, continental climate).

An example of a structural thermal bridge, balcony slab, is shown in the left side of Fig. 3.18 and in Fig. 3.19. Insulation is penetrated by the balcony slab, through which the heat flow rate significantly increases. The two possible methods to mitigate this include insulating the external part of the balcony slab (Fig. 3.18, centre) or cantilevering thermal bridge through the load-bearing thermal insulation element, consisting of a special high-tensile steel reinforcement and insulator (Fig. 3.18, right and Fig. 3.20).

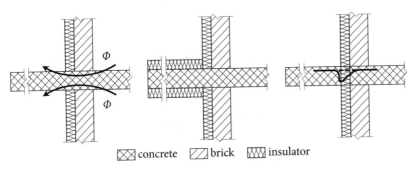

▨ concrete ▨ brick ▨ insulator

Figure 3.18: Sketch of a balcony slab. Because the slab penetrates the insulation, the heat flow rate is increased (left). There are two possibilities for mitigation: insulate the external part of the balcony slab (centre) or use special elements consisting of a thermal insulator pierced by a special high-tensile steel reinforcement that cantilevers the balcony slab (right).

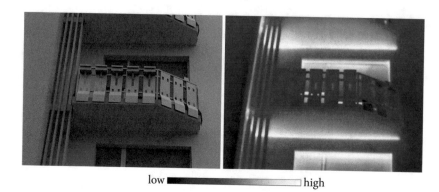

low ▮▬▬▬▬▬▬▬▯ high

Figure 3.19: Thermographic analysis of a balcony slab. The heat flow is increased because the balcony slab interrupts the insulation.

We must also take into account that the thermal bridge often contributes significantly to the heat flow rate. As already pointed out, the density of heat flow changes direction, and the isotherms are not planar, so the exact heat flow rate cannot be calculated analytically. In order to simplify the calculation, we can 'localise' (take into account only the close surroundings of) a particular thermal bridge and then *calculate the difference in heat flow rate* in situations with and without a particular thermal bridge. We then use the result to define *linear thermal transmittance* Ψ and *point thermal transmittance* χ. In the end, new thermal transmittances can be used to calculate the heat flow rate of any thermal bridge of that type. Some calculation details will be elaborated on below, whereas the full procedure is specified in ISO 10211 [39].

Another important issue regarding thermal bridges is reduced internal surface temperature θ_{si}. As shown in Fig. 3.13 on page 90, the temperature is considerably lower in the edge than in the remaining parts of the internal surface. As we will discuss in Section 4.5.3 on page 141, knowledge of this temperature is essential to determine whether the risks of condensation and mould growth exist. In practical situations this temperature can be calculated only from the temperature factor of the internal surface $f_{R_{si}}$ (3.8), where internal surface temperature corresponds to the smallest in-

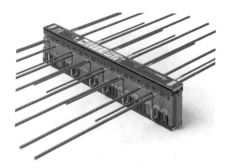

Figure 3.20: Load-bearing thermal insulation element [65]. It consists of a thermal insulator pierced by a special high-tensile steel reinforcement that cantilevers the balcony slab.

ternal surface temperature in the area of the thermal bridge. The temperature factor is determined by numerical calculation of thermal bridges, together with linear or point thermal transmittances.

3.2.3 Dimension systems

Calculating the contribution of thermal bridges to the heat flow rate depends on the dimension system. The three dimension systems defined in ISO 13789 [45] and depicted in Fig. 3.21 are

1. *external dimensions*, measured between the finished external faces of the external building elements of the building;

2. *internal dimensions*, measured between the finished internal faces of each room in a building; and

3. *overall internal dimensions*, measured between the finished internal faces of the external building elements of the building.

In this book, we will address only the internal and external dimensions.

Let's consider the effect of different dimension systems on a simple thermal bridge—a homogeneous building edge (Fig. 3.13 on page 90, Table 3.2). For an approximation, we can assume that the complete heat flow rate is simply the heat flow rate through two planes that are connected by the thermal bridge. However, as is evident from the sketch, in cases of external dimensions for the external edge and internal dimensions for the internal edge, the heat flow through the protruding part is calculated twice. We therefore expect that approximate heat flow rate Φ_{ap} is larger than exact heat flow rate Φ_{ex}, meaning that the correction for the heat flow rate, described by linear thermal transmittance Ψ, is negative. On the other hand, in cases of external dimensions for the internal edge and internal dimensions for the external edge, the heat flow through the protruding part is not calculated at all. We therefore expect that the approximate heat flow rate is smaller than the exact heat flow rate, meaning that the correction for the heat flow rate, described by linear thermal transmittance Ψ, is positive. Thus, linear thermal transmittance Ψ can be either positive or negative, depending on the dimension system and the type of thermal bridge.

> **Info box**
>
> Thermal bridge heat flow rate contribution can be either positive or negative.

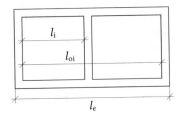

Figure 3.21: Dimension systems shown in a floor plan of a building with two rooms. Length l_e represents the external dimension, l_{oi} represents the overall internal dimension and l_i represents the internal dimension.

Table 3.2: Effect of different dimension systems for a homogeneous building edge of length/height l. Φ_{ap} is the approximate heat flow rate and Φ_{ex} is the exact heat flow rate. The linear thermal transmittance can be either positive or negative.

	External Dimensions	Internal Dimensions
External Edge	$\Phi_{ap} = l\, l_{e1} U_1 \Delta\theta + l\, l_{e2} U_2 \Delta\theta$ $\Phi_{ex} < \Phi_{ap} \implies \Psi < 0$	$\Phi_{ap} = l\, l_{i1} U_1 \Delta\theta + l\, l_{i2} U_2 \Delta\theta$ $\Phi_{ex} > \Phi_{ap} \implies \Psi > 0$
Internal Edge	$\Phi_{ap} = l\, l_{e1} U_1 \Delta\theta + l\, l_{e2} U_2 \Delta\theta$ $\Phi_{ex} > \Phi_{ap} \implies \Psi > 0$	$\Phi_{ap} = l\, l_{i1} U_1 \Delta\theta + l\, l_{i2} U_2 \Delta\theta$ $\Phi_{ex} < \Phi_{ap} \implies \Psi < 0$

3.2.4 Linear thermal bridges

We will first demonstrate the calculation of the linear thermal bridge's heat flow contribution as specified by ISO 10211 [39]. For demonstration purposes, we use a homogeneous building edge (Fig. 3.13 on page 90), but the same procedure can be used for any linear thermal bridge. The problem is two-dimensional because heat does not flow along the thermal bridge. It is reasonable to assume that exact heat flow rate Φ_{ex} through the edge and its surroundings is proportional to the temperature difference $\Delta\theta$ and the edge length l (Fig. 3.22, left):

$$\Phi_{ex} = l\, L_{2D} \Delta\theta.$$

The quantity L_{2D} $(\mathrm{W/(m\,K)})$ is called the *thermal coupling coefficient from two-dimensional numerical calculation* [39]. On the other hand, although the heat flows through the planes and their contacts are intertwined (Fig. 3.13 on page 90), the total heat flow rate can be conceived as the sum of heat flow rates through adjoined planes, Φ_1, Φ_2, and some additional heat flow rate Φ (Fig. 3.22, right):

$$\Phi_{ex} = \Phi_1 + \Phi_2 + \Phi.$$

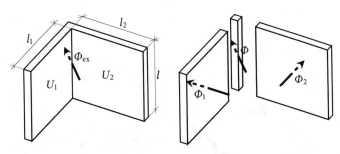

Figure 3.22: Linear thermal bridge. The exact heat flow rate Φ_{ex} (left) can be conceived as the sum of heat flow rates through adjoined planes, Φ_1, Φ_2, and some additional heat flow rate Φ (right).

Assuming that the additional heat flow rate is also proportional to the temperature difference $\Delta\theta$ and the edge length l, we can finally write

$$\Phi_{\text{ex}} = l\, l_1 U_1 \Delta\theta + l\, l_2 U_2 \Delta\theta + l\, \Psi \Delta\theta.$$

Joining the first and the last equation, we finally get

$$\Psi = L_{\text{2D}} - l_1 U_1 - l_2 U_2.$$

The quantity Ψ $(\text{W}/(\text{m K}))$ is called the *linear thermal transmittance* and multiplied by the length of the thermal bridge and temperature difference gives the thermal bridge heat flow rate contribution as

$$\Phi = l\, \Psi \Delta\theta. \tag{3.34}$$

Standard ISO 10211 [39] provides further details on creating a computational model for linear thermal bridges, which are, however, beyond the scope of this book.

3.2.5 Point thermal bridges

Now we will demonstrate how to calculate the heat flow rate contribution of a point thermal bridge, as specified by ISO 10211 [39]. For demonstration purposes, we use a homogeneous building corner, but the same procedure can be used for any point thermal bridge. The problem is three-dimensional, as heat flows in all directions. It is reasonable to assume that exact heat flow rate Φ_{ex} through the corner and its surroundings is proportional to the temperature difference $\Delta\theta$ (Fig. 3.23, left):

$$\Phi_{\text{ex}} = L_{\text{3D}} \Delta\theta.$$

The quantity L_{3D} (W/K) is called the *thermal coupling coefficient from three-dimensional numerical calculation* [39]. On the other hand, although the heat flows through the planes and their contacts are intertwined, the total heat flow can be conceived as the sum of heat flow rates through

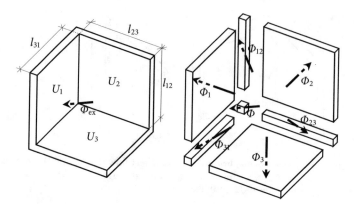

Figure 3.23: Point thermal bridge. The exact heat flow rate Φ_{ex} (left) can be conceived as the sum of the heat flow rates through adjoined planes, Φ_1, Φ_2, Φ_3; adjoined linear thermal bridges, $\Phi_{12}, \Phi_{23}, \Phi_{31}$; and some additional heat flow rate Φ (right).

adjoined planes, Φ_1, Φ_2, Φ_3; through adjoined linear thermal bridges, $\Phi_{12}, \Phi_{23}, \Phi_{31}$; and some additional heat flow rate Φ (Fig. 3.23, right):

$$\Phi_{ex} = \Phi_1 + \Phi_2 + \Phi_3 + \Phi_{12} + \Phi_{23} + \Phi_{31} + \Phi.$$

Assuming that the additional heat flow rate is also proportional to the temperature difference $\Delta\theta$, we can finally write

$$\Phi_{ex} = l_{12}l_{31}U_1\Delta\theta + l_{12}l_{23}U_2\Delta\theta + l_{23}l_{31}U_3\Delta\theta +$$
$$+ l_{12}\Psi_{12}\Delta\theta + l_{23}\Psi_{23}\Delta\theta + l_{31}\Psi_{31}\Delta\theta + \chi\Delta\theta.$$

Joining the first and the last equation, we get

$$\chi = L_{3D} - l_{12}l_{31}U_1 - l_{12}l_{23}U_2 - l_{23}l_{31}U_3 - l_{12}\Psi_{12} - l_{23}\Psi_{23} - l_{31}\Psi_{31}.$$

Noting that three linear thermal transmittances have to be calculated beforehand as

$$\Psi_{12} = L_{2D,12} - l_{31}U_1 - l_{23}U_2,$$
$$\Psi_{23} = L_{2D,23} - l_{12}U_2 - l_{31}U_3,$$
$$\Psi_{31} = L_{2D,31} - l_{12}U_1 - l_{23}U_3,$$

we finally obtain

$$\chi = L_{3D} - l_{12}L_{2D,12} - l_{23}L_{2D,23} - l_{31}L_{2D,31} + l_{12}l_{31}U_1 + l_{12}l_{23}U_2 + l_{23}l_{31}U_3.$$

The quantity χ (W/K) is called the *point thermal transmittance* and multiplied by the temperature difference, it gives the thermal bridge heat flow rate contribution as

$$\Phi = \chi\Delta\theta. \tag{3.35}$$

According to ISO 14683 [46], the influence of point thermal bridges, insofar as they result from the intersection of linear thermal bridges, can be generally neglected.

Standard ISO 10211 [39] provides further details on creating a computational model for point thermal bridges, which is, however, beyond the scope of this book.

3.2.6 Thermal transmittance of windows

We have already addressed thermal transmittance of glazing in Section 3.1.4. However, glazing has to be surrounded by the frame that gives it support and enables the window to be opened. The complete heat flow rate through a simple window is therefore the sum of the heat flow rate through glazing Φ_g, the heat flow rate through the frame Φ_f and the heat flow rate through the junction between the frame and glazing Φ_ψ as

$$\Phi = \Phi_g + \Phi_f + \Phi_\psi =$$
$$= A_g U_g \Delta\theta + A_f U_f \Delta\theta + l_g \Psi_g \Delta\theta,$$

where A_g and U_g are the area and thermal transmittance of glazing, A_f and U_f are the area and thermal transmittance of the frame and l_g and Ψ_g are the length and linear thermal transmittance of the junction between the frame and glazing. If we define thermal transmittance of window U_W as

$$\Phi \equiv A_W U_W \Delta\theta = (A_f + A_g) U_W \Delta\theta,$$

we obtain

$$U_W = \frac{A_g U_g + A_f U_f + l_g \Psi_g}{A_f + A_g}. \tag{3.36}$$

To keep the window thermal transmittance low, the frame thermal transmittance has to be addressed as well. The choice of material dictates the profile of the frame. Timber is a relatively good thermal insulator, so the profile of the frame can be full. Plastic and metal are good conductors, however. In order to reduce thermal transmittance, the frame has to contain air chambers (see Fig. 3.24).

More precise instructions on calculating the thermal transmittance of windows can be found in ISO 10077-1 [35].

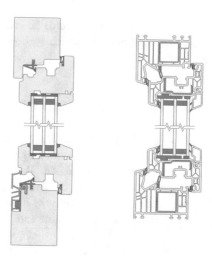

Figure 3.24: Timber window frame (left) and polyvinyl chloride (PVC) window frame (right) for triple-glazed window. Due to higher thermal conductivity, the vinyl profile is not full but contains air chambers in order to increase the frame's thermal resistance.

3.3 Overall heat losses of the building

One of the fundamental tasks in building physics is maintaining a comfortable temperature within a building. Because internal temperature is generally different from external temperature, the heat flow rate between the interior and exterior is unavoidable. These heat losses (or gains) have to be replaced in order to preserve the internal temperature (see Section 1.8 on page 19), so we have to know the value of the overall heat flow rate.

We start by defining basic concepts. *Conditioned space* is part of the building that is heated and/or cooled to maintain a specified temperature. *Unconditioned space* is part of the building that generally does not have its own heat sources and/or sinks, but its temperature differs from the external temperature due to thermal contact with the conditioned space. The building as a whole is in thermal contact with the external environment and the ground. These two entities should be discriminated because, as we have already explained in Section 2.2.6 on page 43, the temperature of the ground differs from the external temperature. Finally, the building's *thermal envelope* is usually defined as the limits of the conditioned space, that is, the separation between the conditioned space on one side and the unconditioned space, ground and external environment on the other side. The thermal envelope is denoted as a dashed line in Fig. 3.25.

In this section we will apply knowledge from previous sections of this chapter and also consider other contributions to heat losses. The four principal contributions are (Fig. 3.25):

1. *direct heat losses*, that is, heat transfer between the conditioned space and external environment;

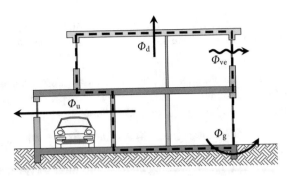

Figure 3.25: Heat losses in a building consist of direct heat losses Φ_d, heat losses through the ground Φ_g, heat losses through unconditioned spaces Φ_u and heat losses through ventilation Φ_{ve}. Densely dashed lines denote building's thermal envelope.

2. *heat losses through the ground,* that is, heat transfer through the ground;

3. *heat losses through unconditioned spaces;*

4. *heat losses through ventilation,* that is, heat transfer due to air ventilation.

Standard ISO 13789 [45] also considers heat losses to adjacent buildings, which are not elaborated in this book.

3.3.1 Direct heat losses

In order to calculate the heat flow rate between the conditioned space and the external environment, we must sum the heat flow rates through all intermediate building components (3.4), the heat flow rates through all linear thermal bridges (3.34) and the heat flow rates through all point thermal bridges (3.35):

$$\Phi_d = \sum_j A_j U_j (\theta_i - \theta_e) + \sum_k l_k \Psi_k (\theta_i - \theta_e) + \sum_l \chi_l (\theta_i - \theta_e).$$

It is usual to define *direct heat transfer coefficient* H_d (W/K) as the quotient of the complete heat flow rate between the conditioned space and external environment Φ_d by the difference between internal and external temperatures:

$$\Phi_d = H_d (\theta_i - \theta_e).$$

Taking two equations together, we see that the direct heat transfer coefficient is only a function of the building properties, independent of internal and external temperatures:

$$H_d = \sum_j A_j U_j + \sum_k l_k \Psi_k + \sum_l \chi_l. \tag{3.37}$$

In order to obtain a reliable result, we need values for linear and point thermal transmittances. There are four methods to determine these values [46]:

1. Numerical calculations. The linear thermal transmittance is calculated in accordance with ISO 10211 [39]. The procedure is outlined in Sections 3.2.4 and 3.2.5. Typical accuracy is ±5 %.

2. Thermal bridge catalogues. The values for a particular thermal bridge can be picked from the catalogue and provided by a third party (for example, a building material producer). Typical accuracy is ±20 %.

3. Manual calculations. Typical accuracy is ±20 %.

4. Default values. ISO 14683 [46] provides default values for most typical thermal bridges calculated for parameters representing worst-case scenarios. These values can be used in the absence of more appropriate data for the concerned thermal bridges. Typical accuracy is from 0 % to 50 %.

The standard ISO 6946 [27] also specifies the calculation of heat losses through pitched roofs:

1. Roof structure consisting of a flat, insulated ceiling and a pitched roof, with naturally ventilated roof spaces. The roof space can be considered as a thermally homogeneous layer with a thermal resistance that depends on the properties of the roof.

2. Heated space under an insulated pitched roof.

   - The values for horizontal heat flow apply to heat flow directions ±30° from the horizontal plane.

   - The values for convective surface coefficients can be obtained by linear interpolation between horizontal and vertical values. Note that radiative surface coefficient is independent of the slope.

3.3.2 Heat losses through the ground

The heat flow rate through the thermal envelope in contact with the ground is calculated separately, using standard ISO 13370 [42]. The standard considers three types of the underlying building part:

1. *Slab on the ground* corresponds to a floor construction directly on the ground over its whole area.

2. *Suspended floor* corresponds to a floor construction in which the lowest floor is held off the ground, resulting in an air void between the floor and the ground.

3. *Basement* corresponds to the usable part of a building that is situated partly or entirely below ground level. Calculations differ for *conditioned* and *unconditioned* basements.

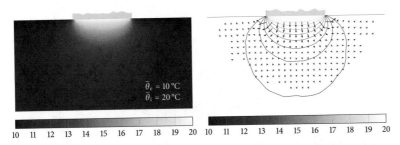

Figure 3.26: Temperatures (left), isotherms and density of heat flow rate vectors (right) for heat transfer through the ground. A stationary situation with average year temperatures is assumed. Results were obtained using FreeFem++ [64].

The plane that separates the calculation of direct heat losses and heat losses through the ground is

- at the level of the inside floor surface for slab-on-ground floors, suspended floors and unconditioned basements, or

- at the level of the outside ground surface for conditioned basements.

Calculating ground heat losses is particularly complex for several reasons:

- The temperature of the ground differs from the temperature of the external air. More on ground temperature time variation (without the presence of the building) can be found in Section 2.2.6 on page 43.

- The heat flow rate within the ground is not homogeneous but three-dimensional. A ground cross-section for temperatures, isotherms and density of heat flow rate vectors for average year temperatures is shown in Fig. 3.26. Numerical calculations are necessary to determine the temperatures and the total heat flow rate leaving the house through the ground.

- The ground region under the thermal influence of the building is typically several times larger than the dimensions of the building itself (Fig. 3.26).

Despite these facts, the annual average heat flow rate through contact between building and ground $\overline{\Phi}_g$ can be calculated as a function of annual average internal $\overline{\theta}_i$ and external $\overline{\theta}_e$ temperatures only. We can therefore define *ground heat transfer coefficient* H_g (W/K) as

$$\overline{\Phi}_g = H_g(\overline{\theta}_i - \overline{\theta}_e).$$

The ground heat transfer coefficient for slab-on-ground and suspended floors corresponds to expression (3.37) as

$$H_g = A\,U + P\,\Psi_g, \tag{3.38}$$

where A is the area of floor, U is its thermal transmittance, P is the exposed perimeter of floor and Ψ_g is linear thermal transmittance associated

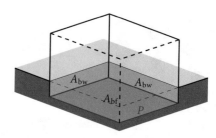

Figure 3.27: Conditioned basement: The area of the basement floor A_{bf}, the area of
the (external) basement walls A_{bw} and the exposed perimeter of the
floor P are the main geometric parameters for calculating heat losses
through the ground.

with the wall/floor junction. The exposed perimeter of the floor is defined
as the total length of the external wall dividing the heated building from
the external environment or from an unheated space outside the insulated
fabric. As pointed out in Section 3.3.1, Ψ_g can be obtained by several methods.

On the other hand, the ground heat transfer coefficient for the conditioned
basement is

$$H_g = A_{bf}\, U_{bf} + A_{bw}\, U_{bw} + P\, \Psi_g, \tag{3.39}$$

where A_{bf} is the area of the basement floor, U_{bf} is its thermal transmittance,
A_{bw} is the area of the external basement walls or basement walls towards
unheated rooms and U_{bw} is their thermal transmittance (Fig. 3.27).

Note that due to the previously mentioned complications, U, U_{bf} and
U_{bw} cannot be easily determined by expressions (3.3) or (3.13). The detailed
calculation is much more complex and specified by ISO 13370 [42].

As shown on the right side of Fig. 3.26, the density of heat flow rate is
largest at the edges of the floor. This means that additional insulation in
this region can significantly reduce the heat flow rate through the ground.
Fig. 3.28 shows three possible edge insulations. For these situations, standard ISO 13370 prescribes calculation of another (negative) linear thermal
transmittance $\Psi_{g,e}$ that takes into account the additional insulation.

Calculating month-dependent heat flow rates through the ground is a very
complex task. We have to take into consideration that the temperature of
the ground differs from the temperature of the external air and that the

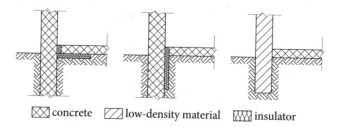

$\boxtimes$ concrete $\diagup\diagup$ low-density material $\bowtie$ insulator

Figure 3.28: Horizontal edge insulation (left), vertical edge insulation (middle)
and low-density foundation (right).

period of temperature change is 12 months (Section 2.2.6 on page 43). The expression (assuming constant internal temperature) is transformed to

$$\Phi_g = H_g(\bar{\theta}_i - \bar{\theta}_e) + H_{pe}\,\Delta\theta_e \cos\left(\frac{2\pi}{12}(t - t_0 - \alpha)\right),$$

where H_{pe} is the external periodic heat transfer coefficient, t_0 is the month number in which the minimum external temperature occurs and α is the time lag of the heat flow cycle compared with that of the external temperature, in months. H_{pe} and α can be obtained by numerical modelling or by the approximate calculation described in ISO 13370.

3.3.3 Heat losses through unconditioned spaces

The procedure for determination of heat losses through unconditioned spaces is described in ISO 13789 [45]. In order to calculate the heat flow rate between a conditioned space and an external environment through the unconditioned space, we must first calculate

- the direct heat transfer coefficient between the conditioned space and the unconditioned space H_{iu} and

- the direct heat transfer coefficient between the unconditioned space and the external environment H_{ue}.

Both H_{iu} and H_{ue} include the transmission heat transfer calculated from (3.48) as well as the ventilation heat transfer calculated from (3.44):

$$H_{iu} = H_{tr,iu} + H_{ve,iu},$$
$$H_{ue} = H_{tr,ue} + H_{ve,ue}.$$

Using these values, we can write the heat flow rate between the conditioned space and the unconditioned space Φ_{iu} and between the unconditioned space and the external environment Φ_{ue}

$$\Phi_{iu} = H_{iu}(\theta_i - \theta_u),$$
$$\Phi_{ue} = H_{ue}(\theta_u - \theta_e),$$

where θ_u is the temperature of unconditioned space. Because we are considering the *steady heat transfer* with time-independent temperatures, the heat flow rate entering the unconditioned space must be equal to the heat flow rate leaving the unconditioned space as

$$\Phi_{iu} = \Phi_{ue} = \Phi_u,$$

which both are equal to the heat flow rate due through unconditioned spaces Φ_u. By solving the preceding three equations, we get

$$\theta_u = \frac{\theta_i H_{iu} + \theta_e H_{ue}}{H_{iu} + H_{ue}},$$

$$\Phi_u = \frac{H_{iu} H_{ue}}{H_{iu} + H_{ue}}(\theta_i - \theta_e).$$

It is usual to define the *heat transfer coefficient through unconditioned spaces* H_u (W/K) as the quotient of the heat flow rate through unconditioned spaces by the difference between internal and external temperatures:

$$\Phi_u = H_u(\theta_i - \theta_e).$$

Taking two equations together, we see that the heat transfer coefficient through unconditioned spaces is only a function of the building and ventilation properties, independent of internal and external temperatures:

$$H_u = \frac{H_{iu} \, H_{ue}}{H_{iu} + H_{ue}}. \tag{3.40}$$

The procedure for calculating the air flow rates regarding unconditioned spaces is described in ISO 13789 [45].

3.3.4 Heat losses through ventilation

The procedure for determining heat losses through ventilation is described in ISO 13789 [45]. Due to ventilation, the internal air of volume V_a at temperature θ_i is replaced by external air of the same volume at temperature θ_e. The external air must be heated to the internal temperature; hence, the required heat is (1.20)

$$Q = m \, c_p \, (\theta_i - \theta_e) = V_a \, \rho \, c_p \, (\theta_i - \theta_e),$$

where m is mass, $c_p = 1.0 \times 10^3$ J/(kg K) specific heat capacity at constant pressure and $\rho = 1.2$ kg/m^3 density of air. If we divide the required heat by the time in which the air is replaced, we get the heat flow rate due to ventilation Φ_{ve}

$$\Phi_{ve} = \frac{Q}{t} = \frac{V_a}{t} \rho \, c_p \, (\theta_i - \theta_e)$$

$$\Phi_{ve} = q_V \rho \, c_p \, (\theta_i - \theta_e), \tag{3.41}$$

where q_V (m^3/s or m^3/h)

$$q_V = \frac{V_a}{t} \tag{3.42}$$

is the *air flow rate*. Note here that the air flow rate is closely related to the *air change rate* n (1/s or 1/h), which is the quotient of the air flow rate by the total volume of the building V

$$n = \frac{q_V}{V}, \tag{3.43}$$

and which essentially indicates how many times is the air within the building completely exchanged within the given period.

It is usual to define *ventilation heat transfer coefficient* H_{ve} (W/K) as the quotient of the heat flow rate due to ventilation by the difference between internal and external temperatures:

$$\Phi_{ve} = H_{ve}(\theta_i - \theta_e).$$

Taking the equations together, we see that the ventilation heat transfer coefficient is only a function of the ventilation and air properties, independent of internal and external temperatures:

$$H_{ve} = \rho\, c_p\, q_V. \qquad (3.44)$$

Airtightness

Apart from obvious reasons, such as mechanical air exhausts (range hood), ducts (chimneys), vents and airing (natural air exchange through window openings), there are always ventilation heat losses through leaks in the building envelope. This contribution is usually described either by the building *airtightness*, that is, the resistance of the building to air leakage through *unintentional* leakage points or areas in the building envelope, or by the opposite building *air permeability*. Typically, leakage occurs at junctions between walls, floors, roof, window frames and door frames, and through electrical and other installations. Leaks in the building envelope contribute significantly to the total heat losses, must be rigorously controlled and are given special consideration in the passive house concept (Section 3.4).

Quantitative methods were developed to calculate or measure the air permeability of buildings. The calculation of the air flow rate q_V is described for example in EN 16798-7 [18] and ISO 52019-2 [51].

On the other hand, the air permeability of buildings can be measured by the fan pressurization method defined by the standard ISO 9972 [34]. This method is usually performed by a device called a blower door, which is an assembly mounted on the door and contains a fan or blower that creates a controlled air flow rate in or out of the building, as shown in Fig. 3.29. Due to the air flow rate through the fan, the air pressure inside the building rises above or below the air pressure outside. At a constant pressure difference, the air flow rate through the blower door is equal to the air flow rate through the building envelope, which is called the *air leakage rate at the reference pressure difference* q_{pr} (m³/s or m³/h). Similarly to air flow rate, the air leakage rate is the quotient of the volume of air transferred through the building envelope V_a by time in which the air is transferred

$$q_{pr} = \frac{V_a}{t}. \qquad (3.45)$$

The typical pressure difference used is 50 Pa and the corresponding *air leakage rate at 50 Pa* is referred to as q_{50}.

The air leakage is closely related to the *air change rate at the reference pressure difference* n_{pr} (1/s or 1/h), which is the quotient of the air leakage rate at the reference pressure difference by the total volume of the building V

$$n_{pr} = \frac{q_{pr}}{V}. \qquad (3.46)$$

For a typical pressure difference of 50 Pa the corresponding *air change rate at 50 Pa* is referred to as n_{50}. Alternatively, the air changes per hour (ACH)

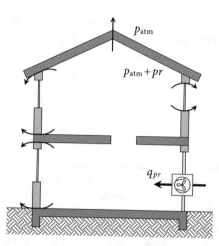

Figure 3.29: Setup for the blower door measurement of the whole building. Here, blower door generates a pressure difference pr by a controlled air flow rate in the building q_{pr}. The latter is compensated by air leakage through leakage points or areas in the building envelope.

is often used as a term for this quantity. A less commonly used quantities are the specific leakage rates as the quotient of the air flow rate by the total envelope area or of the air flow rate by the total floor area.

Note that units for air flow rate, air leakage rate and air change rate are usually given *per hour*, although basic SI units are given *per second*. In this case the specific heat capacity c_p in (3.41) must be given in $W\,h/(kg\,K)$. Upper limits for the air change rate at 50 Pa are usually prescribed by standards [7] or national legislation.

Measurements of the air permeability of buildings can also be used for a more reliable calculation of air flow rate q_V. In its simplest form, the air change rate at 50 Pa n_{50} and the air change rate (in the absence of pressure difference) n can be related by an empirical equation

$$n = \frac{n_{50}}{N},\tag{3.47}$$

where N is the leakage-infiltration ratio. Similarly, the air flow rate q_V can be determined from the air leakage rate at 50 Pa q_{50}

$$q_V = \frac{q_{50}}{N}.$$

The comprehensive model for North America [66] provided the leakage-infiltration ratio for various climates and also took into account the height, shielding and leakage of the building. The combination of the leakage-infiltration ratio and all corrections gives a range of values between 6 and 44. On the other hand, the standard ISO 13789 [45] prescribes a value $N = 20$.

Currently, in Northern America building air permeability is measured according to the standard ASTM E779 [6], with a conventional reference pressure of 4 Pa. ASHRAE 62.2 [5] then prescribes a more accurate determination of air flow rate, taking into account the vertical distance between the lowest and highest above-grade points within the pressure boundary, and the weather and shielding factor, which is specified for more than a thousand US and Canadian locations.

3.3.5 Transmission and overall heat losses

The transmission heat flow rate is the heat flow rate due to direct heat losses, heat losses through the ground and heat losses through unconditioned spaces:

$$\Phi_{tr} = \Phi_d + \Phi_g + \Phi_u,$$

It is usual to define the *transmission heat transfer coefficient* H_{tr} (W/K) as the quotient of transmission heat flow rate Φ_{tr} by the difference between internal and external temperatures [45]:

$$\Phi_{tr} = H_{tr}(\theta_i - \theta_e).$$

The transmission heat transfer coefficient is therefore the sum of the previously defined heat transfer coefficients:

$$H_{tr} = H_d + H_g + H_u. \tag{3.48}$$

Note that the overall heat losses also include heat losses through ventilation, so the total flow rate of heat losses is

$$\Phi = \Phi_{tr} + \Phi_{ve} = (H_{tr} + H_{ve})(\theta_i - \theta_e). \tag{3.49}$$

3.4 Passive house

Passive house is a rigorous, voluntary standard for energy-efficient building:

> A passive house is a building, for which thermal comfort (ISO 7730) can be achieved solely by post-heating or post-cooling of the fresh air mass, which is required to achieve sufficient indoor air quality conditions – without the need for additional recirculation of air. [76]

In a passive house, the conventional means of heating and cooling (for example, furnaces, central heating or air conditioners), which heat or cool the internal air, are not allowed. Regardless of that, a ventilation of minimum air flow rate $q_V = 30 \, \text{m}^3/\text{h} = 8.3 \times 10^{-3} \, \text{m}^3/\text{s}$ per person is required to maintain a reasonable indoor air quality. The standard allows for fresh air to be heated or cooled, so the building is essentially conditioned by ventilation only.

To achieve thermal comfort, several conditions must be met, which can be roughly divided into four categories.

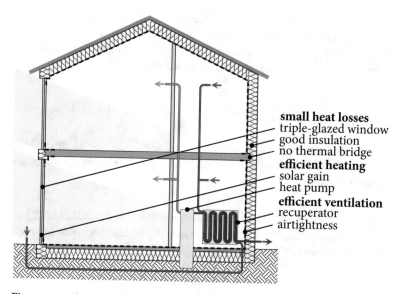

Figure 3.30: The passive house design. To achieve low energy requirements for the building energy balance, several conditions must be met. These conditions can be roughly divided into four categories: small heat losses, efficient heating, efficient ventilation and shading (not shown).

1. *Small heat losses.* Direct heat losses, heat losses through the ground and heat losses through unconditioned spaces are sufficiently reduced under these conditions (Fig. 3.30):

   - A low thermal transmittance of all nontransparent building components should be achieved using good thermal insulation.

   - Triple-glazed windows of thermal transmittance $U_W \leq 0.8\,\mathrm{W/(m^2\,K)}$ [77] should be used.

   - Thermal bridges should be avoided or their effects minimised.

2. *Efficient ventilation.* In order to reduce heat losses through ventilation, we need to address two issues (Fig. 3.30):

   - The air flow rate should be reduced by increasing the *airtightness* of the building (Section 3.3.4). This is ensured by the passive house requirement that the air change rate at 50 Pa should be less than 0.6/h [77], which corresponds approximately to a natural air change rate of 0.03/h (3.47).

   - Regardless of this, as already mentioned, ventilation of minimum air flow rate $q_V = 30\,\mathrm{m^3/h} = 8.3 \times 10^{-3}\,\mathrm{m^3/s}$ per person is necessary to maintain adequate indoor air quality, and corresponding heat losses are unavoidable. If the dwelling has $30\,\mathrm{m^2}$ of living space per person with a ceiling height of 2.5 m, this results in an air change rate of 0.4/h, which is well above the permissible natural air change rate. To reduce ventilation losses, passive houses use forced ventilation through the *recuperator* (Section 2.3.2 on page 52).

3. *Efficient warming.* It is impossible to avoid heat losses completely, therefore a certain amount of external energy is necessary to maintain thermal comfort (Fig. 3.30):

- The standard permits post-heating of fresh air as a means of heating. To increase post-heating efficiency, a *heat pump* must be used.

- In many climates, heating through ventilation will not be sufficient, so that part of the heating is provided by *solar gain* (Section 3.1.5 on page 85), in particular energy harvested by exploiting the *greenhouse effect* (Section 2.4.1 on page 56). This means that passive houses must have *large, equator-oriented transparent building components* (glass surfaces) of solar factor $g \geq 0.5$ [77].

4. *Shading.* On the other hand, large solar gain can be very disadvantageous in warmer parts of the world and for warmer seasons, when the interior of the building needs to be cooled rather than heated. In these cases, shading systems with a low solar factor must be used.

With the preceding assumptions, it is possible to calculate the upper limit of energy consumption independently of climatic conditions. The fresh air optionally flows first through a ground-coupled heat exchanger (underground pipes) and then through the recuperator to increase its temperature almost to the internal temperature. The air is then additionally heated for a maximum of $\Delta\theta = 30\,°C$ to avoid pyrolysis of dust, which starts at about $50\,°C$. Using minimum air flow rate of $q_V = 8.3 \times 10^{-3}\,m^3/s$ and (3.41), we see that the maximum permissible heat consumption is $\Phi_V = 300\,W$ per person. The general assumption is that the dwelling has $30\,m^2$ of living space per person, which gives a more practical maximum permissible heat consumption of $10\,W/m^2$. Furthermore, for a continental climate, the annual heating energy should not exceed $15\,kW\,h/m^2$ of the net living space (treated floor area) [77].

Problems

3.1 Vertical glass of thickness 6 mm separates internal the environment at 20 °C and the external environment at −1 °C. Calculate the temperatures at the surfaces of the glass. Take the thermal conductivity of the glass to be 0.80 W/(m K). (3.7 °C, 4.6 °C)

3.2 Vertical double glazing with glasses of thickness 4.0 mm, thermal conductivity of 0.80 W/(m K) and airspace of thickness 16.0 mm between the glasses, separates the internal space at 20 °C from the external environment. Take the thermal resistance of the airspace to be 0.19 m² K/W, and calculate external temperature for which the temperature of the internal surface of the double glazing is 15 °C. Calculate the other characteristic temperatures. (5.8 °C; 7.3 °C, 7.5 °C, 14.8 °C)

3.3 Vertical walls of the thermal envelope consist of a concrete layer of thickness 20.0 cm and thermal conductivity 1.0 W/(m K) and a façade stone of thickness 2.0 cm and thermal conductivity 2.0 W/(m K). Calculate the thermal transmittance of the uninsulated and insulated walls, where the latter is obtained by adding an extra layer of expanded polystyrene (EPS) of thickness 15.0 cm and thermal conductivity 0.035 W/(m K). For the external temperature −5 °C and internal temperature 20 °C, calculate the characteristic temperatures for all three situations (|stone|concrete|, |stone|concrete|EPS|, |stone|EPS|concrete|). Plot the temperature as a function of layer thickness and as a function of thermal resistance. (2.63 W/(m² K), 0.214 W/(m² K); −2.4 °C, −1.7 °C, 11.5 °C; −4.8 °C, −4.7 °C, −3.7 °C, 19.3 °C; −4.8 °C, −4.7 °C, 18.2 °C, 19.3 °C)

3.4 The total area of a room's thermal envelope is 50 m²: The vertical windows of thermal transmittance 1.1 W/(m² K) make up 20 m² of the envelope, and three wall types from problem 3.3 make up the rest of the area. Assume that the external and internal temperatures are constants with values −5 °C and 20 °C, respectively, and that the heat value of the natural gas (amount of heat released during the combustion per cubic meter of gas) is about 33 MJ/m³. Calculate the daily gas volume consumption for heating the room for uninsulated and insulated walls. What are the savings (percentage) for the insulated wall? Is this percentage dependent on the internal and external temperatures? (6.6 m³, 1.9 m³; 72 %; no)

3.5 Vertical airspace of thickness 5.0 mm is enclosed by surfaces of emittance 0.90. Calculate the convective and radiative surface coefficient if the thermal resistance of the airspace is 0.11 m² K/W. Take the average temperature to be 275 K. (3.9 W/(m² K), 5.2 W/(m² K))

3.6 Calculate the external surface resistance for the façade with emittance 0.50. Take the average of external and surface temperatures to be 10 °C and the average wind speed to be 10 m/s. (0.021 m² K/W)

3.7 The vertical walls of a thermal envelope consists of

- a façade brick of thickness 100 mm and thermal conductivity 0.76 W/(m K);
- an airspace of thickness 50 mm, on one side enclosed with aluminium foil;
- a fibre cement board of thickness 16 mm and thermal conductivity 0.35 W/(m K);
- mineral wool of thickness 150 mm and thermal conductivity 0.04 W/(m K); and
- a plasterboard of thickness 15 mm and thermal conductivity 0.21 W/(m K).

Take the convective surface coefficient for airspace to be $1.25\,\text{W}/(\text{m}^2\,\text{K})$ and emittances of the surfaces to be 0.90 and 0.05. Assume that the average temperature within the airspace is $0\,°\text{C}$. Calculate the thermal transmittance of the wall, characteristic temperatures and temperature factor of the internal surface if internal and external temperatures are $21\,°\text{C}$ and $-5\,°\text{C}$, respectively. $(0.206\,\text{W}/\text{m}^2;\ -4.8\,°\text{C},\ -4.1\,°\text{C},\ -0.5\,°\text{C},\ -0.2\,°\text{C},\ 19.9\,°\text{C},\ 20.3\,°\text{C},\ 0.97)$

3.8 The vertical timber frame construction substructure, the floor plan of which is shown in the figure, has two rows of vertical timber beams of dimensions $60\,\text{mm} \times 60\,\text{mm}$ and thermal conductivity $0.14\,\text{W}/(\text{m}\,\text{K})$ separated for 565 mm along the wall and for 40 mm perpendicular to the wall. The substructure is closed by plasterboard of thickness 25 mm and thermal conductivity of $0.21\,\text{W}/(\text{m}\,\text{K})$ on one side, and with fibre cement board of thickness 16 mm and thermal conductivity $0.35\,\text{W}/(\text{m}\,\text{K})$ plastered with mortar of thickness 8 mm and thermal conductivity $0.50\,\text{W}/(\text{m}\,\text{K})$ on the other side. The intermediate space is filled with mineral wool of thermal conductivity $0.040\,\text{W}/(\text{m}\,\text{K})$. Calculate the upper limit and lower limit of the total thermal resistance and thermal transmittance of the wall using the simplified calculation. $(3.98\,\text{m}^2\,\text{K}/\text{W},\ 3.77\,\text{m}^2\,\text{K}/\text{W},\ 0.258\,\text{W}/(\text{m}^2\,\text{K}))$

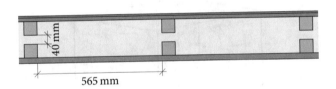

40 mm

565 mm

3.9 A window with dimensions $150\,\text{cm} \times 100\,\text{cm}$ consists of 10 cm wide frame of thermal transmittance $1.4\,\text{W}/(\text{m}^2\,\text{K})$ and the glazing of thermal transmittance $1.1\,\text{W}/(\text{m}^2\,\text{K})$. The linear thermal transmittance on the junction of the frame and glazing is $0.070\,\text{W}/(\text{m}\,\text{K})$. Calculate the thermal transmittance of the entire window. $(1.39\,\text{W}/(\text{m}^2\,\text{K}))$

3.10 A room of height $h = 2.5\,\text{m}$ has two external mutually perpendicular walls of lengths $a = 4.0\,\text{m}$ and $b = 7.0\,\text{m}$, as shown in the figure. The wider wall has two windows of height 1.5 m and width 2.0 m. Take the thermal transmittance of the walls to be $0.25\,\text{W}/(\text{m}^2\,\text{K})$, the thermal transmittance of the windows to be $1.1\,\text{W}/(\text{m}^2\,\text{K})$, the linear thermal transmittance at the junction of the two walls to be $-0.030\,\text{W}/(\text{m}\,\text{K})$ and the linear thermal transmittance at the junction of the window and wall to be $0.050\,\text{W}/(\text{m}\,\text{K})$. Calculate the required power for heating the room, assuming internal and external temperatures to be $20\,°\text{C}$ and $-10\,°\text{C}$, respectively. $(378\,\text{W})$

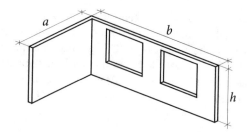

3.11 Two-storey house in shape of cuboid of width a = 10.0 m, depth b = 8.0 m and height h = 6.0 m with a flat roof is erected on the ground, as shown in the figure. The house has windows of total area 30 m². Take the thermal transmittance of the roof to be 0.25 W/(m² K), the thermal transmittance of the vertical walls to be 0.21 W/(m² K), the thermal transmittance of the windows to be 1.1 W/(m² K), the linear thermal transmittance at the junction of the vertical walls to be −0.050 W/(m K) and the linear thermal transmittance at the junction of the roof and vertical walls to be 0.20 W/(m K). The remaining thermal bridges can be neglected. Calculate the direct heat transfer coefficient. Calculate the heat flow rate due to direct heat losses if the internal and external temperatures are 21 °C and 5 °C, respectively. (98.1 W/K; 1570 W)

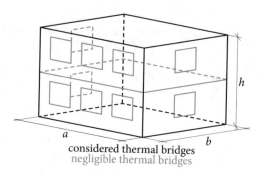

considered thermal bridges
negligible thermal bridges

4 Moisture in building components

In this chapter we will first list the causes of excessive moisture and the problems it generates for a building and its occupants. To study the transfer of water between the environment and the building, we will discuss how to describe water contents and then examine the transfer mechanisms. Finally, we will use the presented knowledge to establish measures to prevent the causes of excessive moisture.

4.1 Introduction

Water as a source of life is ever present in our environment. In certain situations, however, it can cause serious problems. In this chapter, we will consider one such situation—*excessive moisture in building components*. Excessive moisture causes the following five problems:

1. *Deteriorated habitation quality.* Increased moisture in building components and consequently increased air humidity boosts the growth of microorganisms, such as bacteria or mould. This has a negative impact on occupants' health as well as surface aesthetics.

2. *Reduced thermal resistance.* Because water has high thermal conductivity, its presence in some materials, particularly in thermal insulators, can significantly increase the thermal conductivity of the material.

3. *Additional mechanical stresses.* Damped materials, especially timber, are prone to expansion. Unaccounted expansion of some components causes additional mechanical stresses and can compromise the stability of the building.

4. *Salt transport.* Liquid water can dissolve salts and carry them as it flows through the building element. Some of the dissolved salts are inherent to the material, whereas others are brought from the outside by rain water or ground water. Salts are then deposited

Figure 4.1: Concrete decay due to corrosion (rust stains) and decalcification (white stains and stalactites).

wherever the liquid flow is disturbed or ended, most notably by water evaporation. Salt crystallisation on the surface, called *efflorescence*, is usually harmless but leaves unaesthetic (usually white) stains. However, if salts crystallise within pores or on boundaries between materials, this causes additional mechanical stresses and breaks the material apart, leading to cracks in building elements or to chipping of cover layers (plaster, paint).

5. *Material decay.* The presence of water causes degradation of building materials. For reinforced concrete, water induces oxidation of reinforcement iron (*corrosion*) and washes calcium from concrete (*decalcification*). These processes are usually observed as rust and white stains on the surface (Fig. 4.1). On the other hand, with timber, water destroys the material's cellular structure. These effects can compromise the integrity of the building element.

In order to describe moisture build-up, we will study the four physical mechanisms that are connected to it, which are

1. *condensation,*

2. *hygroscopy (sorption),*

3. *capillary action* and

4. *diffusion.*

Excessive moisture is due to both obvious as well as more inconspicuous causes (Fig. 4.2):

A *Moisture intrusion into building component due to contact with liquid water* is the most obvious cause. The most common reasons for this are

- the building being in contact with damp ground,

- precipitation (rain, snow) falling onto the building or

- leaks in installations (water, sewage) and in the roof.

By *capillary action*, water is then absorbed deep into the material, even against gravity.

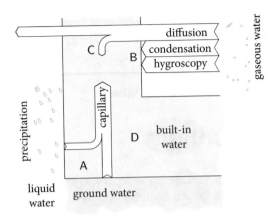

Figure 4.2: Causes of excessive moisture.

B *Moisture deposition on the building component surface due to contact with gaseous water* is another important cause. There are two physical mechanisms that lead to water transfer from the air to the building component surface: *condensation* and *hygroscopy (sorption)*. Note that moisture deposed on the surface can be transferred further into the material due to capillary action.

C *Moisture intrusion into the building component due to contact with gaseous water* is also an important but less conspicuous cause. Water is transferred directly into the building component due to *diffusion* and is deposited there by condensation.

D *Built-in moisture* is the result of certain building procedures. Chemical processes in concrete and plaster casting produce excess water that can remain within the material years after casting. Fresh timber also contains large quantities of water.

We will describe these mechanisms and causes in more detail in the rest of this chapter.

In this chapter, we will come across the mass transfer of water. In short time periods, the amount of transferred water is practically always proportional to time: Water mass transferred in 1 min will be 60 times larger than water mass transferred in 1 s. It is therefore convenient to define *mass flow rate* q_m (kg/s) as the quotient of transferred mass by time:

$$q_m = \frac{dm}{dt}. \tag{4.1}$$

For a stationary situation, that is, a time-independent water flow, a nondifferential form can be used as

$$q_m = \frac{m}{t}. \tag{4.2}$$

Generally, the water flow rate is proportional to the area, so it is usually convenient to define *density of water (vapour) flow rate* $g\,(\mathrm{kg/(m^2\,s)})$ as

$$q_m = A\,g.\qquad(4.3)$$

4.2 Air humidity

As already pointed out, moisture in building components is closely related to water vapour in the air. Air humidity is therefore an essential concern in building physics.

We start our study by considering the equilibrium between liquid and gaseous water. As we pointed out in Section 1.7 on page 16, in a phase diagram equilibrium is represented by the line that separates liquid or solid from gaseous water (Fig. 1.11 on page 18). We called this line *water vapour pressure at saturation* p_{sat}. We can read out values for two distinguished equilibrium points, $\theta = 20\,°\mathrm{C}$, $p_{\mathrm{sat}} = 2.3\,\mathrm{kPa}$, and $\theta = 100\,°\mathrm{C}$, $p_{\mathrm{sat}} = 101.3\,\mathrm{kPa}$.

Water vapour pressure at saturation is determined experimentally and can be looked up in tables or calculated with quasi-empirical equations. ISO 13788 [44] specifies useful equations, that will be used throughout the rest of this book:

$$p_{\mathrm{sat}} = 610.5 \exp\left(\frac{17.269\,\theta}{237.3 + \theta}\right),\ \theta \geq 0\,°\mathrm{C},$$

$$p_{\mathrm{sat}} = 610.5 \exp\left(\frac{21.875\,\theta}{265.5 + \theta}\right),\ \theta < 0\,°\mathrm{C}.\qquad(4.4)$$

The inverted equations are

$$\theta = \frac{237.3\ln\left(\frac{p_{\mathrm{sat}}}{610.5}\right)}{17.269 - \ln\left(\frac{p_{\mathrm{sat}}}{610.5}\right)},\ p_{\mathrm{sat}} \geq 610.5\,\mathrm{Pa},$$

$$\theta = \frac{265.5\ln\left(\frac{p_{\mathrm{sat}}}{610.5}\right)}{21.875 - \ln\left(\frac{p_{\mathrm{sat}}}{610.5}\right)},\ p_{\mathrm{sat}} < 610.5\,\mathrm{Pa}.\qquad(4.5)$$

Understanding equilibrium is important because phase transitions, conversion to vapour (vaporisation, sublimation) and conversion from vapour (condensation, deposition) occur in the equilibrium state. We are most familiar with vaporisation. To boil water in a pot at atmospheric pressure, it must be warmed to $100\,°\mathrm{C}$. On the other hand, from our experience, we know that water condensation also takes place near room temperatures, such as water drops on window glazing or dew droplets on plants. How is it possible for liquid and gaseous water to be in an equilibrium state at room temperature?

The answer is that we have take into account only the partial pressure of water and not the pressure of the whole air. We have already pointed out

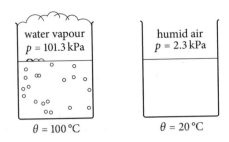

Figure 4.3: Partial pressure of water vapour for boiling (left) and for room temperature evaporation (right). In both cases, the partial pressure of the water vapour is equal to the saturated one. For the water at room temperature, the partial pressure of the dry air fills in the difference to full atmospheric pressure.

that the pressure of the gas mixture is the sum of the partial pressures of its components (Section 1.4.1 on page 9). For humid air, we can write (1.13)

$$p_{atm} = p(N_2) + p(O_2) + p(Ar) + \cdots + p(H_2O) = p_d + p, \qquad (4.6)$$

where $p_d = p(N_2) + p(O_2) + p(Ar) + \ldots$ is the partial pressure of dry air, and $p = p(H_2O)$ is the partial pressure of water vapour.

There are two types of *vaporisation*:

1. For water at 100 °C (Fig. 4.3, left), the phase transition occurs within the bulk of the water (bubbles), and the created water vapour displaces all the air above the water surface. Partial pressure of water vapour is equal to atmospheric pressure $p = 101.3\,kPa$, whereas the partial pressure of the dry air is zero $p_d = 0.0\,kPa$. The process is called *boiling*.

2. For water at room temperature (Fig. 4.3, right), the phase transition occurs only on the surface of the water, and just above the surface, all air components are present. The partial pressure of the water vapour is less than atmospheric pressure $p = 2.3\,kPa$, whereas the partial pressure of dry air $p_d = 99.0\,kPa$ fills in the difference to full atmospheric pressure. The process is called *evaporation*.

In both cases, liquid and gaseous water are in equilibrium because saturation is related to partial water vapour pressure and not to the total pressure of humid air.

Nature tends to establish equilibrium state, so when there is a surplus of water vapour (Fig. 4.4, left), condensation occurs. This process is extremely quick, as the system releases energy, so normally we don't encounter partial water vapour pressure larger than that at saturation. On the other hand, when there is a lack of water vapour (Fig. 4.4, right), vaporisation occurs. The process is slow, as the system requires additional energy. Furthermore, when there is an absence of liquid water, equilibrium cannot be achieved at all.

Therefore, note that in practical situations, the upper limit of water vapour pressure is equal to that at saturation. It is therefore suitable to define their

> **Info box**
>
> To study humid air, we split it into dry air and water vapour.

> **Info box**
>
> There are two types of vaporisation, one that occurs in the bulk is called boiling, the other that occurs at the surface is called evaporation.

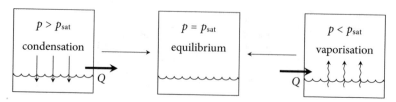

Figure 4.4: The state with a surplus of water vapour (left) and the state with a lack of water vapour (right) tend to establish equilibrium (centre). The process of reducing the water vapour amount (condensation) is extremely quick, as the system releases energy (heat). The process of increasing the water vapour amount (vaporisation) is slow, as the system requires additional energy.

ratio called *relative humidity* φ as

$$\varphi = \frac{p}{p_{sat}}$$ (4.7)

Info box

In practical terms, water vapour at saturation describes the air capacity for water vapour.

and express it in percentages. For $\varphi = 100\%$, we say that air is saturated, so that the water vapour quantity achieved the maximum capacity of air. For $\varphi < 100\%$, air is unsaturated and can accept more water vapour.

In Fig. 4.5, water vapour pressures are presented for different relative humidities. Relative humidity can be increased by increasing the water vapour amount and hence increasing water vapour pressure (dashed arrow). In that case, we do not change the capacity of air, but the relative humidity increases due to the larger water vapour amount. Relative humidity can also be increased by decreasing the temperature and water vapour pressure at saturation (full arrow). In that case, we do not change the water vapour amount but *reduce the capacity of air*, again leading to the same result. In contrast, by decreasing the water vapour pressure or increasing the temperature, relative humidity is decreased.

Info box

By cooling the air, we reduce its capacity for water vapour and thus increase relative humidity.

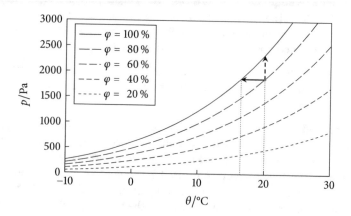

Figure 4.5: Water vapour pressure for different relative humidities. Relative humidity can be increased by increasing the water vapour amount and hence increasing the water vapour pressure (dashed arrow), by decreasing the temperature and water vapour pressure at saturation (full arrow), or a combination thereof. The opposite applies for decreasing relative humidity.

Figure 4.6: The inhaled air is warmed and moistened inside the lungs. As this warm, humid air is exhaled into a low-temperature environment, the air quickly cools down, decreasing the water vapour pressure at saturation and increasing the relative humidity. This leads to water vapour saturation and condensation of water, which is visible as a mist trail.

The effect of condensation due to the decreased temperature and decreased water vapour pressure at saturation can be observed in many common situations, such as human exhalation at low temperatures (Fig. 4.6) or chimney exhausts. Notorious contrails or 'chemtrails' also share the same effect. Plane engine exhaust contains large amounts of carbon dioxide and water vapour as a result of fuel combustion. Exhaust gas quickly cools at high altitudes, leading to condensation of water that is visible as condensation or vapour trails.

The temperature to which the air must be cooled to become saturated and trigger condensation is called the *dew point*. Dew point can be estimated from Fig. 4.5. Let's assume that the air temperature is 20 °C and the relative humidity is 80 %. If the air is cooled to 16.5 °C near the wall surface, the air becomes saturated. If there is a further temperature decrease, $\varphi > \varphi_{\text{sat}}$, condensation occurs. In order to prevent condensation of water vapour on the walls, their surface temperature should be kept high.

Another common physical quantity for describing water vapour content is the quotient of water vapour mass m by volume of gas V, *mass concentration of water vapour v* (kg/m^3):

$$v = \frac{m}{V}. \tag{4.8}$$

This quantity is more commonly called *absolute humidity*. In the rest of the book, we will use the expression prescribed by the standard.

Using ideal gas law (1.12), the definition can be reformulated in terms of water vapour pressure p as

$$v = \frac{p}{R_v T}, \tag{4.9}$$

where *gas constant for water vapour R_v* is the quotient of the molar gas constant by the molar mass of water $M_v = 0.018\,\text{kg/mol}$ as

$$R_v = \frac{R}{M_v} = 461.5\,\frac{\text{J}}{\text{kg K}}. \tag{4.10}$$

The third common physical quantity for describing water vapour content is the ratio of water vapour mass m to the mass of dry gas m_d, *mass ratio of water vapour to dry gas x*:

$$x = \frac{m}{m_d}.$$ (4.11)

This quantity is more commonly called the *humidity ratio*. In the rest of the book, we will use the abbreviated expression prescribed by the standard, that is, water vapour mass ratio. It has no units but is commonly displayed in g/kg.

Using (1.12) and (4.6), the definition can be reformulated in terms of water vapour pressure p as

$$x = \frac{M_v}{M_d}\frac{p}{p_d} = c\,\frac{p}{p_{atm} - p},$$

where $M_d = 0.029\,kg/mol$ is the molar mass of dry air, and

$$c = \frac{M_v}{M_d} = 0.622.$$

4.3 Psychrometrics

Psychrometrics is the field of engineering concerned with the determining physical and thermodynamic properties of gas-vapour mixtures. So far, we have used temperature and amount of water vapour to describe humid air properties. In psychrometrics, we are also interested in humid air energy. For this purpose, we will use enthalpy as defined in Section 1.6 on page 13.

Similar to potential energy, the absolute value of enthalpy has little significance as we are primarily interested in the quantity changes. So we are free to choose a reference point, and in psychrometrics the most convenient reference point corresponds to liquid water of mass m and dry air of mass m_d at temperature $\theta = 0\,°C$. Let's calculate enthalpy of humid air at temperature θ. We have to vaporise all water and then increase the temperature of both the water vapour and dry air to θ (1.28,1.32):

$$H = m\,h_v + (m\,c + m_d\,c_d)\theta.$$

In this equation, $h_v = 2.501 \times 10^6\,J/kg$ is the specific enthalpy of vaporisation at $0\,°C$, and $c = 1926\,J/(kg\,K)$ and $c_d = 1005\,J/(kg\,K)$ are the specific heat capacities at constant pressure for water vapour and dry air, respectively.

In psychrometrics, we are especially interested in *specific enthalpy h*, that is the quotient of enthalpy by mass of dry air,

$$h = \frac{H}{m_d},$$ (4.12)

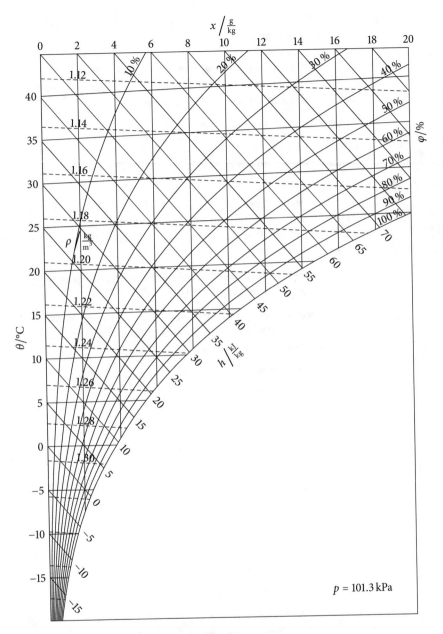

Figure 4.7: Mollier diagram.

which for humid air is equal to

$$h = x\,h_{\mathrm{v}} + (x\,c + c_{\mathrm{d}})\theta.$$

For a selected pressure of humid air (usually atmospheric pressure 1.013×10^5 Pa), all essential psychrometric quantities can be conveniently presented in the same *psychrometric chart*. The most common psychrometric charts are *Mollier diagrams* (Fig. 4.7) and ASHRAE-style psychrometric charts. Both charts include specific enthalpy h, water vapour mass ratio x, relative humidity φ and temperature θ (in latter called the dry bulb temperature). For convenience, many psychrometric charts, such as Fig. 4.7, include the density of air ρ. ASHRAE-style psychrometric chart also includes the wet bulb temperature, which is the temperature air would have if it was cooled to saturation by the vaporisation of water into it, with the latent heat being supplied by the air.

Psychrometric charts have many uses:

- When we know the humid air state, that is, two psychrometric quantities, we can easily look up the values of the remaining quantities.

- We can easily inspect various humid air state changes. When humid air undergoes various processes, all psychrometric quantities can change, *but one quantity is always constant—the mass of the dry air m_d.* Additionally,

  - water vapour mass ratio x is also constant for processes with constant water vapour content and

  - specific enthalpy h is also constant for adiabatic processes, that is, processes in which humid air does not exchange heat with the environment (see Section 1.9.2 on page 23).

- We can easily obtain the properties of humid air, which is a result of mixing two amounts of humid airs.

Example 4.1: Cooling the humid air.

The room contains $m_a = 50\,\text{kg}$ of humid air at temperature $\theta_1 = 25\,°C$ and relative humidity $\varphi_1 = 70\,\%$.

1. Determine the water vapour mass ratio and specific enthalpy of the humid air. Calculate the mass of dry air, enthalpy of the air and mass of water vapour.

2. An air conditioner cools the air to temperature $\theta_2 = 10\,°C$. Determine the final relative humidity. Calculate the energy that is removed and the water mass that is extracted from the air in the process.

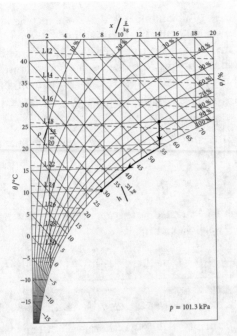

1. The initial state of the air is located at the intersection of the $\theta = 25\,°C$ line and the $\varphi = 70\,\%$ line. At this point, we can read out $x_1 = 14.0\,g/kg$ and $h_1 = 61\,kJ/kg$.

   We can calculate the mass of dry air by noting that the mass of humid air is the sum of the mass of dry air and water vapour mass. Combining that with expression (4.11), we get

   $$m_a = m_d + m = m_d + x_1\, m_d$$

   $$\implies m_d = \frac{m_a}{1 + x_1} = 49.3\,kg. \qquad (4.13)$$

   Next, we get enthalpy from (4.12) as

   $$H = h_1\, m_d = 3.01\,MJ$$

   and mass of water vapour from (4.11) as

   $$m_1 = x_1\, m_d = 0.69\,kg.$$

2. As pointed out previously, the mass of dry air remains constant. When we cool humid air, we subtract energy, which reduces the enthalpy of the system. Furthermore, as long as the air is not saturated, the quantity of the water vapour remains constant. Hence, from (4.11), we conclude that the water vapour mass rate must also be constant, so we proceed along the x-line.

   Because the temperature and consequently the water vapour pressure at saturation decrease, the relative humidity increases. This continues to the dew point when relative humidity is 100 % and the air is saturated. From that moment on, we can no longer proceed along the x-line because we would end up in a 'forbidden' zone of oversaturated air. We therefore proceed along $\varphi = 100\%$ (saturation) line.

The final state of the air is characterised by $\varphi_2 = 100\%$, $x_2 = 7.6\,\text{g/kg}$ and $h_2 = 29\,\text{kJ/kg}$.

To get removed energy Q, we subtract the final and initial enthalpy:

$$Q = \Delta H = H_2 - H_1 = m_\text{d}\, h_2 - m_\text{d}\, h_1 = -1.58\,\text{MJ}.$$

To get the extracted water vapour, we subtract the final and initial water vapour mass:

$$\Delta m = m_2 - m_1 = m_\text{d}\, x_2 - m_\text{d}\, x_1 = -0.32\,\text{kg}.$$

Example 4.2: Evaporation humidifier.

A room contains $m_\text{a} = 100\,\text{kg}$ of humid air at temperature $\theta_1 = 25\,°\text{C}$ and relative humidity $\varphi_1 = 30\%$. With the help of an evaporative humidifier, we adiabatically increase relative humidity to $\varphi_2 = 60\%$. Determine the final temperature of the air and the mass of the evaporated water.

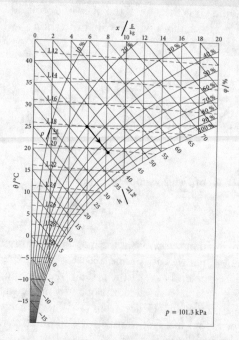

$p = 101.3\,\text{kPa}$

From the Mollier diagram, we can figure out that the initial state of the air is characterised by $x_1 = 5.9\,\text{g/kg}$ and $h = 40\,\text{kJ/kg}$. The principle of the evaporative humidifier is to increase the water surface area by creating little water droplets. This way, evaporation is accelerated without adding any heat to the system. This process therefore follows the h-line.

The final state of the air is located on the intersection of the $h = 40\,\text{kJ/kg}$ line and the $\varphi = 60\,\%$ line. At this point, we can read out $x_2 = 8.2\,\text{g/kg}$ and $\theta_2 = 18.5\,°\text{C}$.

To calculate the mass of evaporated water, we must first calculate the mass of dry air (4.13) as

$$m_d = \frac{m_a}{1 + x_1} = 99.4\,\text{kg}.$$

To get the mass of evaporated water, we subtract the final and initial water vapour mass:

$$\Delta m = m_2 - m_1 = m_d\,x_2 - m_d\,x_1 = 0.23\,\text{kg}.$$

What is happening during the evaporation process? Because no energy is added to the system, the energy is simply rearranged within the system (liquid water and humid air). Humid air is cooled down in order to give away energy, which is then used to evaporate the water.

4.4 Building components moisture

The majority of building materials are porous, which means that they contain voids:

porous materials	nonporous materials
concrete	glass
gypsum plaster	steel
brick	ceramic, porcelain
timber	polyvinyl chloride (PVC)
thermal insulators	

Despite the fact that these voids are mostly microscopic and invisible to the naked eye, their volume and surface are far from negligible. The quantity of porous space is often described by the *specific internal surface area* as the quotient of total air void surface by total mass of the sample. For example, the specific internal surface area is about $0.2\,\text{m}^2/\text{g}$ for gypsum plaster and about $20\,\text{m}^2/\text{g}$ for cement paste [78]. As a result, building materials can attract, contain and transfer considerable amounts of water.

Info box

Because of porosity, building materials can attract, contain and transfer water.

The amount of contained water (water vapour as well as liquid water) is described by several physical quantities. The most common quantity is *mass ratio of water to dry matter u* as in

$$u = \frac{m_w}{m_d},$$ (4.14)

where m_w is the mass of water, and m_d is the mass of dry matter. Note that it differs from the water vapour mass ratio (4.11) in that the former counts in the complete mass of water vapour and liquid water, whereas the latter counts in only the mass of water vapour. This quantity is customarily called *water (moisture) content*. In the rest of the book, we will use the standard expression in abbreviated form as water mass ratio.

The water mass ratio can be determined easily. The sample is weighted in its damp state to get the total mass of the sample m_{tot}, put into the oven to dry and weighted again to get m_d. The water mass ratio is then

$$u = \frac{m_{tot} - m_d}{m_d}.$$

Another common physical quantity usually used for low-density materials is *mass concentration of water w* (kg/m^3) as in

$$w = \frac{m_w}{V},$$ (4.15)

where m_w is the mass of water, and V is the volume of the sample. Note that it differs from the mass concentration of water vapour (4.8) in that the former counts in the complete mass of water vapour and liquid water, whereas the latter counts in only the mass of water vapour.

Finally, water content is also quantified by *moisture content volume by volume ψ* [40] as in

$$\psi = \frac{V_w}{V},$$ (4.16)

where V_w is the volume of water, and V is the volume of the sample. Note that both mass ratio of water to dry matter u and moisture content volume by volume ψ are dimensionless, but they have different values for the same water content.

As mentioned in Section 4.1, the humidity of building materials is increased by four mechanisms: condensation, hygroscopy, capillary action and diffusion. We have already elaborated on condensation, so in the following sections we will take a closer look at the other three mechanisms.

4.4.1 Hygroscopic damping

We have already studied condensation, that is, the process in which liquid water is extracted from the air due to saturation. However, another physical process extracts water from the air even if the air is not saturated, that is, for relative humidities $\varphi < 100\,\%$.

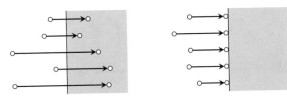

Figure 4.8: Types of sorption. In absorption (left), ions and molecules of one substance enter a substance in a different state of matter. In adsorption (right), ions and molecules bond onto the surface of a substance in a different state of matter.

Hygroscopy is the ability of a substance to attract and hold water molecules from the surrounding environment. The processes involved are somewhat different and more complex than condensation and vaporisation described previously. They are based on the electromagnetic interaction between the polar water molecules and the polar molecules of the material.

> **Info box**
>
> Due to hygroscopy, materials can get damp even if the air is not saturated, that is, without condensation.

In terms of hygroscopy, materials can be generally divided into two categories:

- *Hygroscopic materials* are damped in contact with humid air and are characterised by large porosity, that is, a large specific internal surface area, which water molecules are attached to. The most common hygroscopic building materials are concrete, gypsum plaster and timber.

- *Nonhygroscopic materials* are not damped in contact with humid air. The most common nonhygroscopic building materials are glass, steel, brick, thermal insulators and polyvinyl chloride (PVC).

Hygroscopic damping is the consequence of the *sorption* mechanism. Sorption is a physical and chemical process by which one substance becomes attached to another. There are two specific types of sorption:

- *Absorption* is the incorporation of ions and molecules of one substance into another substance in a different state of matter (Fig. 4.8, left).

- *Adsorption* is the adherence or bonding of ions and molecules of one substance onto the surface of another substance in a different state of matter (Fig. 4.8, right).

Desorpion is an opposite mechanism of sorption, in which a substance is released from or through a surface.

Regarding building materials, the amount of contained water is regulated by adsorption and desorption. As it turns out, the amount of adsorbed water is closely connected to the humidity of the air. This relation is usually depicted by a *sorption (moisture) isotherm*, that is, water mass ratio or mass concentration of water plotted against relative humidity for a fixed temperature (Fig. 4.9).

As shown in Fig. 4.9, isotherms for adsorption, or material damping, and desorption, or material drying, differ. Let's assume that we want to achieve water mass ratio $u = 0.1$. For damping, we have to put the material into

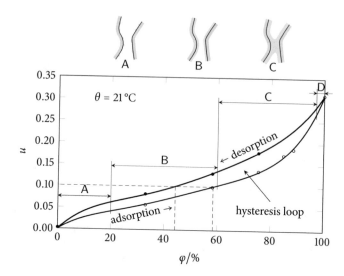

Figure 4.9: Sorption isotherm for timber (sugar maple) [67]. The water mass ratio depends not only on relative humidity but also on the underlying process. Such behaviour is called hysteresis. Four basic damping regions (A to D) are described in the text.

air with relative humidity $\varphi = 58\,\%$ for long period of time. On the other hand, for drying, we have to put the material into air with relative humidity $\varphi = 45\,\%$ for a long period of time. The response of the system (mass water ratio) is therefore dependent not only on current conditions (relative humidity), but also on its history. Such dependence is called *hysteresis*.

As demonstrated in Fig. 4.9, there are four basic damping regimes:

A Adsorbed molecules in the material form a single layer. The approximate relative humidity range is $\varphi < 20\,\%$.

B Adsorbed molecules in the material form multiple layers. The approximate relative humidity range is $60\,\% > \varphi > 20\,\%$.

C Layers of adsorbed molecules are interconnected. The approximate relative humidity range is $97\,\% > \varphi > 60\,\%$.

D Pores are completely filled with water, and capillary suction steps in. The approximate relative humidity range is $\varphi > 97\,\%$.

Hygroscopy is also related to salt (NaCl) damage of building materials. Common salt is highly hygroscopic and attracts additional water into the pore structure of concrete. This process leaves less room for expansion in the pore structure, which creates more pressure inside the concrete when it freezes.

Finally, hygroscopy is not necessarily harmful. By adsorption and desorption of water molecules, hygroscopic materials regulate air moisture. By absorption of water during high air humidity and desorption of water during low air humidity, they mitigate humidity fluctuations.

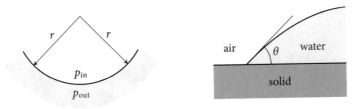

Figure 4.10: Pressures on the outside and inside of the surface sphere (left) and contact angle θ (right).

4.4.2 Capillary damping

Capillary damping is based on the physical principle of surface tension. Liquids possess the tendency to form surfaces of the smallest possible area. This tendency can be easily understood in terms of energy. A molecule in contact with another neighbouring molecule is in a lower state of energy than if it was alone. Whereas interior molecules have a maximum number of neighbour molecules and possess minimal possible energy, boundary molecules are missing neighbour molecules and possess higher energy. In order to minimise the total energy, liquid tends to minimise the number of boundary molecules, which leads to a minimisation of surface area. To describe this tendency, we introduce material-dependent physical quantity *surface tension* γ (N/m). Namely, the force required to increase the surface depends on the properties of liquid and is proportional to the length of the surface boundary.

Surface tension has two consequences:

1. *Laplace pressure* Δp (Pa). In the absence of gravitational force, liquid will form a perfect spherical drop, which represents the minimal surface area. Although there is no gravitational force, attractive forces still exist between molecules. For symmetry reasons, the net force on the internal molecules is zero, whereas the net energy on the surface molecules is nonzero. That leads to a difference of the pressures right above and below the surface called the Laplace pressure. It can be shown that for a sphere of radius r, it amounts to (Fig. 4.10, left)

$$\Delta p = p_{\text{in}} - p_{\text{out}} = \frac{2\gamma}{r}, \qquad (4.17)$$

where p_{in} is the pressure inside the sphere, and p_{out} is the pressure outside the sphere.

2. *Contact angle* θ. Because surface tensions towards air and various solid materials are different, liquid tends to increase the surface towards one substance at the expense of the surface towards the other. This leads to the contact angle, which is defined as an angle between air-liquid and solid-liquid surfaces (Fig. 4.10, right).

In terms of contact angle we divide substances into two categories:

1. *Hydrophobic materials* that repel water and have a contact angle of $\theta > 90°$.

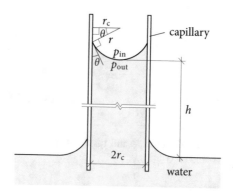

Figure 4.11: Capillary action for a glass capillary immersed into water ($\theta < 90°$). For a small capillary radius, the water surface within the capillary forms almost the perfect spherical shape. Because of the pressure difference, the liquid level within the capillary rises above the level of the remaining liquid.

2. *Hydrophilic materials* that attract water and have a contact angle of $\theta < 90°$.

Note that glass is simultaneously nonhygroscopic and hydrophilic, with a contact angle around 27°.

What happens if a *capillary*—small diameter tube—is immersed into liquid (Fig. 4.11)? The surface of the liquid within the capillary will form almost a perfect spherical shape. Because pressures below and above the spherical surface will be different, the level of liquid in the capillary will rise above, or drop below, the level of the remaining liquid. Using elementary geometry, it is easy to derive the relation between capillary radius r_c and sphere radius r as

$$r_c = r \cos \theta$$

$$\implies p_{in} - p_{out} = \Delta p = \frac{2\gamma \cos \theta}{r_c}.$$

Noting that $p_{in} = p_{atm}$ and using the well-known hydrostatic expression $p_{atm} = p_{out} + \rho g h$, we get

$$h = \frac{2\gamma \cos \theta}{\rho g r_c}, \tag{4.18}$$

where ρ is the density of the liquid. Note that the effect is inversely proportional to the capillary radius.

For hydrophilic materials ($\theta < 90°$), water rises above the level of the remaining water, and for hydrophobic materials ($\theta > 90°$), water drops below that level. The *capillary action* is a very common physical effect responsible for, among other things, water transport in plants.

In civil engineering, capillary action is commonly observed as *rising damp*. Bottom parts of a building are in direct contact with the ground, which is often water soaked. On the other hand, most building materials are hydrophilic and are highly porous, and these pores are interconnected in the

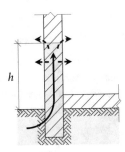

Figure 4.12: Detail of the house explaining rising damp. Ground water enters the porous building element and by capillary action rises to elevation h above the ground level (full arrow). Water is continuously evaporating from the building element (dashed arrows) but is also concomitantly replaced by more ground water, establishing a continuous upward capillary water flow.

capillary structure. If we let the ground water enter the capillary structure, it will rise high above the ground level due to the capillary effect (Fig. 4.12). Water then evaporates from the building element, depositing the dissolved salts in the process. Evaporated water is replaced by ground water, establishing continuous upward capillary water flow.

The presence of the capillary effect and rising damp can be easily identified by inspecting the façade (Fig. 4.13). The wall is damped up to a certain, almost constant level above ground level, permanent stains (efflorescence) appear or façade plaster and/or paint is partly or entirely flaked off. The damage is usually largest in the zone near the moisture boundary, for two reasons. First, evaporation and therefore salt crystallisation is strongest there, so this zone is usually called the evaporation zone. Furthermore, the height of the moisture boundary varies due to changing meteorological conditions and evaporation rates, so this zone goes through a series of damping and drying cycles. Both processes exert additional internal pressure onto the material.

We can make a rough estimate of water elevation h above ground level. The surface tension of water at room temperature is $\gamma = 0.073\,\text{N/m}$, density $\rho = 1000\,\text{kg/m}^3$ and the contact angle is $\theta \approx 0$ for most building materials.

Figure 4.13: The façade damage due to rising damp appears high above the ground level. Note that the largest damage occurs close to the moisture boundary, which is usually called the evaporation zone.

We can therefore estimate that $h = 14.6$ mm for $r_c = 1$ mm, $h = 146$ mm for $r_c = 0.1$ mm and $h = 1460$ mm for $r_c = 0.01$ mm.

The exact calculation of capillary action is more complex due to volatile capillary diameters and capillary directions. One of the most useful expressions is the empirical *Washburn equation*, which predicts that the volume of liquid V that is absorbed by the sample depends on time t as

$$V = AS\sqrt{t}, \tag{4.19}$$

where A is the horizontal cross-sectional area of the sample that is damp on the bottom side, and $S\,(\text{m/s}^{\frac{1}{2}})$ is *sorptivity*. If we define porosity $f = V/V_0$ as the ratio of the pore volume to the volume of the sample, and note that the volume of the damped part of the sample is equal to $V_0 = Ah$, we can show that damping elevation is equal to

$$h = \frac{S}{f}\sqrt{t}. \tag{4.20}$$

Some authors define term S/f as sorptivity instead of S. Note that in this simplified model, we have assumed that water completely fills up the voids, which is generally not true.

Finally, we can also calculate the density of the water flow rate absorbed by the sample using (4.3) and (4.1):

$$g = \frac{q_m}{A} = \frac{1}{A}\frac{dm}{dt} = \frac{\rho}{A}\frac{dV}{dt} = \frac{\rho S}{2\sqrt{t}}. \tag{4.21}$$

Another important issue is *capillary condensation*. The spherical water surface in a capillary increases the water vapour pressure. Therefore, condensation can appear at a water vapour pressure lower than the water vapour pressure at saturation. This effect is described by the *Kelvin equation*

$$\ln \varphi' = -\frac{\Delta p}{\rho R_v T}, \tag{4.22}$$

where φ' is the relative humidity at which condensation appears, Δp is the Laplace pressure, ρ is the density of liquid water, R_v (4.10) is the gas constant for water vapour and T is the temperature. Taking into account the expression for Laplace pressure (4.17), for relative humidity at which condensation appears, we get

$$\varphi' = \exp\left(-\frac{2\gamma}{\rho R_v T r}\right) \approx \exp\left(-\frac{2\gamma}{\rho R_v T r_c}\right). \tag{4.23}$$

For concrete, the capillary radius can be as small as $r_c \approx 1 \times 10^{-8}$ m, leading to condensation at approximately $\varphi' \approx 90\,\%$.

Note that modelling water transfer that is driven by capillary action will be discussed in Section 4.6.2.

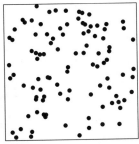

Figure 4.14: The model for the diffusion of black substance particles within a white (invisible) fluid. Initially, all the black substance is located in a small fraction of the container and its concentration is high there, whereas elsewhere the concentration is zero (left). A net movement of black particles from the high concentration zone to the low concentration zone sets in. After longer period of time the concentration of the black substance balances out within the whole container (right).

4.5 Water vapour diffusion

In previous sections, we have learnt that the porosity of the building materials enables the transfer of liquid water. The underlying process is the capillary action. Porosity also enables the transfer of water vapour. As we will see in this section, the underlying process is *water vapour diffusion*.

Diffusion in general is the net movement of anything from a region of higher concentration to a region of lower concentration, as shown in Fig. 4.14. Water vapour diffusion is therefore the net movement of water vapour molecules from a region of higher mass concentration of water vapour to a region of lower mass concentration. In order to study the process, we have to know the water vapour content. The content will be described either by the mass concentration of water vapour itself or the water vapour pressure.

4.5.1 Fick's first law

We start our study of diffusion with *steady diffusion*. Steady implies that we allow for water vapour content variation in space, but no variation in time.

Let's observe water vapour transfer through a slab of material whose opposite faces are at different mass concentrations of water vapour, higher v_h and lower v_l (Fig. 4.15). The slab's cross-sectional area is A, and the thickness is d.

It can be shown experimentally that the water vapour mass flow rate is always

- proportional to cross-sectional area A,
- inversely proportional to thickness d and

> **Info box**
>
> If there is a spatial difference in humidity, the water vapour diffusion from the region with the higher mass concentration of water vapour to the region with the lower mass concentration of water vapour sets in.

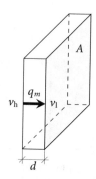

Figure 4.15: Diffusion through a solid slab whose opposite faces are at different mass concentrations of water vapour, $v_h > v_l$. The slab's cross-sectional area is A, and the thickness is d.

- proportional to *mass concentration of water vapour difference* $\Delta v = v_h - v_l$.

We can join these statements into one expression:

$$q_m = D\frac{A\Delta v}{d}.$$ (4.24)

This statement is called *Fick's first law*. The coefficient of proportionality D (m^2/s) is called the *water vapour diffusion coefficient*. Note that the logic of mass transfer as well as the law we have just established is similar to the logic of heat transfer and Fourier's law (2.5).

The mass flow rate also depends on the slab substance. The water diffusion coefficient is thus material dependent and has to be determined experimentally.

In terms of the density of water vapour flow rate (4.3), Fick's first law is transformed to

$$g = D\frac{\Delta v}{d}.$$ (4.25)

Unlike thermal conductivity λ, water vapour diffusion coefficient D temperature dependence cannot be neglected. For example, the most important water vapour diffusion coefficient, the one for stagnant air D_0, can be written to a good approximation with Schirmer equation [41] as

$$D_0 = 2.31 \times 10^{-5} \frac{p_{atm}}{p} \left(\frac{T}{273.15}\right)^{1.81},$$ (4.26)

where p is air pressure.

To eliminate temperature dependence, we describe the water vapour diffusion coefficient for materials in terms of the one for stagnant air. This is done by defining temperature-independent *water vapour resistance factor* μ as the ratio of vapour diffusion coefficients for water vapour for stagnant air D_0 and for arbitrary matter D as

$$\mu = \frac{D_0}{D} \geq 1. \tag{4.27}$$

Because the water vapour diffusion coefficient for air is the largest of the water vapour diffusion coefficients, the water vapour resistance factor is always larger than one. The definition (4.27) transforms Fick's first law into the form

$$g = D_0 \frac{\Delta v}{\mu\, d}. \tag{4.28}$$

The water vapour resistance factor is material dependent and has to be determined experimentally. Values for a few typical building materials are presented in Table A.3 on page 277.

Water vapour barriers are elements that resist water vapour transfer. Note that good water vapour barrier materials have a *large* water vapour resistance factor. Typically, polyethylene sheet and aluminium foil are used for this purpose.

> **Info box**
>
> Good water vapour barrier materials have a large water vapour resistance factor.

Usually, the layer in the building component is characterised by *water vapour diffusion - equivalent air layer thickness* s_d (m), that is, the product of thickness and water vapour resistance factor

$$s_d = \mu\, d. \tag{4.29}$$

As the name suggests, a stagnant air layer of thickness s_d has the same water vapour resistance as a material of thickness d and water vapour resistance factor μ.

Fick's first law in terms of water vapour pressure

Finally, we can express the mass concentration of water vapour in Fick's first law (4.28) in terms of water vapour pressure. Using (4.9), we get

$$g = \delta_0 \frac{\Delta p}{\mu\, d}, \tag{4.30}$$

where

$$\delta_0 = \frac{D_0}{R_v T} \approx 2 \times 10^{-10}\ \frac{\mathrm{kg}}{\mathrm{m\,s\,Pa}} \tag{4.31}$$

is the *water vapour permeability with respect to vapour pressure* of air. Because this value does not change significantly with temperature (water vapour diffusion coefficient for stagnant air D_0 and temperature T increase concomitantly), standard ISO 13788 [44] prescribes the use of the approximated value written above.

> **Info box**
>
> It is more practical to study water vapour diffusion in terms of water vapour pressure.

Table 4.1: Surface equivalent air layer thicknesses in civil engineering [17].

s_d/m	Direction of Water Vapour Flow		
	Upward	Horizontal	Downward
internal, $s_{d,si}$	0.004	0.008	0.03
external, $s_{d,se}$	$\frac{1}{67+90\,v}$	$\frac{1}{67+90\,v}$	$\frac{1}{67+90\,v}$

4.5.2 Mass convection

In Section 2.3.1 on page 49, we argued that in a steady heat transfer situation, only the temperature gradient next to the external building component surface remains. The situation is similar for water vapour.

Due to the of convective movement of air, the mass concentration of water vapour and the water vapour pressure within an environment, such as a room, are also equalised. However, the mass concentration of water vapour and water vapour pressure within an environment differs from the one at the building component surface. In a steady water vapour transfer situation, convection transfers water vapour only between the building component surface and its adjacent environment. We can describe this transfer by a law that is similar to Newton's law of cooling (2.30), (2.31) as

$$q_m = A\,h_m(v_s - v_0),$$
$$g = h_m(v_s - v_0),$$

where A is the surface area, v_0 and v_s are the mass concentrations of water vapour in the environment and at the surface of the wall, respectively. Constant h_m (m/s) is the *surface coefficient of mass transfer* and is generally different for external h_{me} and internal surface h_{mi}. Comparing this equation with (4.28) and (4.29), we can define the *surface equivalent air layer thicknesses* as

$$s_{d,se} = \frac{D_0}{h_{me}}, \quad s_{d,si} = \frac{D_0}{h_{mi}}.$$

According to standard ISO 13788 [44], for practical engineering situations, external and internal surface equivalent air layer thicknesses are independent of building component inclination and are assumed to be

$$s_{d,se} = s_{d,si} = 0.01\,\text{m}. \tag{4.32}$$

More precise surface equivalent air layer thicknesses for civil engineering situations are specified in standard EN 15026 [17]. Its values are shown in Table 4.1. At external surfaces, the thickness can be calculated using the expression

$$\frac{1}{67 + 90\,v},$$

where v (m/s) is the wind speed adjacent to the surface.

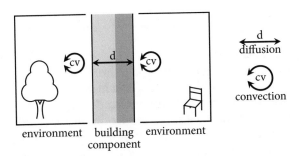

Figure 4.16: Sketch of water vapour transfer through a building component (solid wall). Two environments exchange water vapour by diffusion and convection.

4.5.3 Water vapour diffusion through the building component

For several layers in steady state (constant density of water vapour flow rate g through all layers), the equivalent air layer thickness is the sum of the individual layers:

$$s_{\mathrm{d,eq}} = \sum_i s_{\mathrm{d},i}.$$ (4.33)

Info box

Water vapour transfer in building components combines diffusion and convection, where the latter can be often neglected.

We must also consider water vapour transfers from building component surfaces to the environment, as shown in Fig. 4.16. From the analogy with heat transfer (3.2), the equivalent air layer thickness of the building component, or *total equivalent air layer thickness*, is

$$s_{\mathrm{d,tot}} = s_{\mathrm{d,se}} + \sum_i \mu_i\, d_i + s_{\mathrm{d,si}}.$$ (4.34)

As shown in Example 4.4, in most practical engineering situations, the surface equivalent air layer thicknesses are very small compared to the material equivalent air layer thicknesses. In fact, they are so small, that they even cannot be seen in Fig. 4.18 on page 147. Consequently, water vapour pressures on the surface of the wall and in the environment are practically the same $p_i \approx p_{\mathrm{si}}$, $p_e \approx p_{\mathrm{se}}$. Calculations in standard ISO 13788 [44] therefore strictly assume that

$$s_{\mathrm{d,se}} = s_{\mathrm{d,si}} = 0,$$

$$p_i = p_{\mathrm{si}},$$

$$p_e = p_{\mathrm{se}}.$$

Taking this into account, (4.34) simplifies to

$$s_{\mathrm{d,tot}} = \sum_i \mu_i\, d_i.$$ (4.35)

Temperature factor of the internal surface

The fact that water vapour pressure does not change (significantly) near the surface is in stark contrast to the significant temperature change in the same region: $\theta_e \neq \theta_{se}$ and $\theta_i \neq \theta_{si}$. In winter months, $\theta_i > \theta_{si}$, this results in an increased relative humidity at the surface (Fig. 4.5 on page 122, full arrow). To investigate this problem, we connect the internal surface relative humidity φ_{si} with the internal surface temperature θ_{si}, the internal relative humidity φ_i and the internal temperature θ_i (4.7) using the equation:

$$p_i = p_{si},$$
$$p_{sat}(\theta_i)\,\varphi_i = p_{sat}(\theta_{si})\,\varphi_{si}. \tag{4.36}$$

Note that to determine internal surface temperature θ_{si} computational method from Section 3.1.2 on page 74 is used for homogeneous building components or temperature factor of the internal surface $f_{R_{si}}$ (3.8) for thermal bridges.

This allows us to calculate internal surface relative humidity φ_{si}, first by calculating internal surface temperature θ_{si} and then using the expression:

$$\varphi_{si} = \frac{p_{sat}(\theta_i)}{p_{sat}(\theta_{si})}\,\varphi_i.$$

Often the procedure is the other way round: the *maximum* allowable internal surface relative humidity φ_{si} is given and we have to calculate the *minimum* temperature factor of the internal surface $f_{R_{si}}$. First, using the expression

$$p_{sat}(\theta_{si}) = \frac{\varphi_i}{\varphi_{si}}\,p_{sat}(\theta_i)$$

and the expression (4.5) *minimum* internal surface temperature θ_{si} is calculated, and then the expression (3.8) is used.

Both calculation procedures for thermal bridges are shown graphically below.

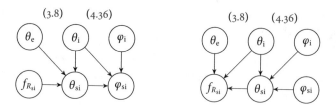

The design of the building elements must ensure that $f_{R_{si}}$ is large enough (3.9) to avoid excessive surface humidity. This is especially critical for thermal bridges, as we have already pointed out in Section 3.2.2 on page 93. Calculation of $f_{R_{si}}$ for thermal bridges is as essential as the calculation of linear Ψ and point χ thermal transmittances.

4.5.4 Determination of characteristic water vapour pressures

If the building component consists only of porous materials, the mass transfer of water vapour between internal and external environments is possible. This transfer has positive and negative consequences: The positive consequence is regulation of internal mass concentration of water vapour, whereas the negative consequence, as we will learn later, is possibility of condensation within building component.

Example 4.3: Water vapour diffusion through the wall.

External walls of area $A = 25\,\text{m}^2$ consist of timber of thickness $d = 20\,\text{cm}$ and of water vapour resistance factor $\mu = 50$. The internal temperature is $\theta_i = 21\,°\text{C}$, and the external temperature is $\theta_e = 4\,°\text{C}$, internal relative humidity is $\varphi_i = 65\,\%$ and external relative humidity is $\varphi_e = 80\,\%$. Calculate the mass concentrations of water vapour in the external and internal environments and the mass flow rate between the environments.

We start by calculating internal and external water vapour pressures using (4.7) and (4.4):

$$p_i = \varphi_i\, p_{sat}(\theta_i) = 65\,\% \times 2486\,\text{Pa} = 1616\,\text{Pa},$$
$$p_e = \varphi_e\, p_{sat}(\theta_e) = 80\,\% \times 813\,\text{Pa} = 650\,\text{Pa}.$$

Note that although the internal relative humidity is smaller than the external relative humidity, the internal water vapour pressure is larger than the external water vapour pressure. This is due to the fact that the internal temperature and internal water vapour pressure at saturation are substantially larger.

Next, we calculate the internal and external mass concentration of water vapour using (4.9):

$$v_i = \frac{p_i}{R_v T_i} = 11.89\,\frac{\text{g}}{\text{m}^3},$$
$$v_e = \frac{p_e}{R_v T_e} = 5.08\,\frac{\text{g}}{\text{m}^3}.$$

Because the internal mass concentration of water vapour is larger than the external one, the water vapour diffusion will be directed from inside to outside. To obtain the mass flow rate, we must first calculate the total equivalent air layer thickness (4.34):

$$s_{d,tot} = s_{d,se} + \mu\,d + s_{d,si} = 10.02\,\text{m}.$$

Finally, we write expression (4.30) for the whole building component and use (4.3) to obtain the mass flow rate:

$$q_m = A\,g = \delta_0\,A\,\frac{p_i - p_e}{s_{d,tot}} = 1.93 \times 10^{-4}\,\frac{\text{g}}{\text{s}} = 16.4\,\frac{\text{g}}{\text{d}}.$$

As shown, the quantity of transferred water vapour is usually very small.

The total equivalent air layer thickness of building component (4.34) depends only on the thicknesses and the water vapour resistance factors of the constituent layers. Most importantly, it is independent of the environment water vapour pressures, which is very convenient.

However, it is often necessary to find characteristic water vapour pressures. Except for the water vapour pressure of internal environment p_i and external environment p_e, we are interested in wall internal surface water vapour pressure p_{si}, wall external surface water vapour pressure p_{se} and interface water vapour pressures (water vapour pressures at the boundaries of two layers) p'. Those water vapour pressures depend on the external and internal water vapour pressures. Here we will introduce two methods for determining water vapour pressures: computational and graphical.

Computational method

Using environment water vapour pressures and the total equivalent air layer thickness, we can calculate the density of water vapour flow rate (4.30):

$$g = \delta_0 \frac{p_i - p_e}{s_{d,tot}}. \tag{4.37}$$

On the other hand, we can use (4.30) to obtain the relationship between water vapour pressures on the opposite boundaries of each layer, where p_h denotes the higher water vapour pressure and p_l is the lower water vapour pressure:

$$g = \delta_0 \frac{p_h - p_l}{\mu d} = \delta_0 \frac{p_h - p_l}{s_d}. \tag{4.38}$$

This equation is also valid for air layers on both sides of the wall. Starting from the internal or external water vapour pressure, we can sequentially calculate all the characteristic water vapour pressures.

Example 4.4: Calculation of characteristic water vapour pressures.

Calculate the typical water vapour pressures for the building component, which is composed of the following layers (the same as in Example 3.1):

Layer	d/m	$\lambda / \frac{W}{m\,K}$	μ
solid layer 1 (façade plate)	0.02	1.5	50
solid layer 2 (EPS)	0.05	0.039	60
solid layer 3 (concrete)	0.15	1	120
solid layer 4 (mortar)	0.02	0.56	25

The internal temperature is $\theta_i = 20\,°C$, the external temperature is $\theta_e = -5\,°C$, the internal relative humidity is $\varphi_i = 40\,\%$ and the external relative humidity is $\varphi_e = 80\,\%$. Take into account the surface equivalent air layer thicknesses.

We start by calculating internal and external water vapour pressures using (4.7) and (4.4):

$$p_e = \varphi_e \, p_{sat}(\theta_e) = 320.9 \, \text{Pa},$$
$$p_i = \varphi_i \, p_{sat}(\theta_i) = 934.8 \, \text{Pa}.$$

Next we calculate the total equivalent layer thickness (4.34) as

$$s_{d,tot} = s_{d,se} + \sum_i \mu_i \, d_i + s_{d,si} = 22.52 \, \text{m}$$

and the density of water vapour flow rate (4.37) as

$$g = \delta_0 \frac{p_i - p_e}{s_{d,tot}} = 5.45 \times 10^{-9} \, \frac{\text{kg}}{\text{m}^2 \, \text{s}}.$$

The density of water vapour flow rate is constant throughout the building component. As shown in Fig. 4.17, we have to find five characteristic water vapour pressures. Let's start from the exterior by writing (4.38) for the external air layer as

$$g = \delta_0 \frac{p_{se} - p_e}{s_{d,se}} \implies p_{se} = p_e + s_{d,se} \frac{g}{\delta_0} = 321.2 \, \text{Pa}.$$

Next, we write (4.38) for the solid layers:

$$g = \delta_0 \frac{p_1' - p_{se}}{\mu_1 d_1} \implies p_1' = p_{se} + \mu_1 d_1 \frac{g}{\delta_0} = 348.5 \, \text{Pa},$$
$$g = \delta_0 \frac{p_2' - p_1'}{\mu_2 d_2} \implies p_2' = p_1' + \mu_2 d_2 \frac{g}{\delta_0} = 430.2 \, \text{Pa},$$
$$g = \delta_0 \frac{p_3' - p_2'}{\mu_3 d_3} \implies p_3' = p_2' + \mu_3 d_3 \frac{g}{\delta_0} = 920.9 \, \text{Pa},$$
$$g = \delta_0 \frac{p_{si}' - p_3'}{\mu_4 d_4} \implies p_{si} = p_3' + \mu_4 d_4 \frac{g}{\delta_0} = 934.5 \, \text{Pa}.$$

For a test, we can write (4.38) for the internal air layer as

$$g = \delta_0 \frac{p_i - p_e}{s_{d,si}} \implies p_i = p_{si} + s_{d,si} \frac{g}{\delta_0} = 934.8 \, \text{Pa}.$$

If previous calculations were precise enough, we should obtain the same internal water vapour pressure as initially specified by the problem.

The calculation of the characteristic water vapour pressures could be started from the interior as well.

Expression (4.38) also can be used to determine the water vapour pressure within the layer, starting from the water vapour pressure on the edge of the layer.

Note that expression (4.37) and multiple expressions (4.38), as written in Example 4.4, can be combined to give an explicit expression for each char-

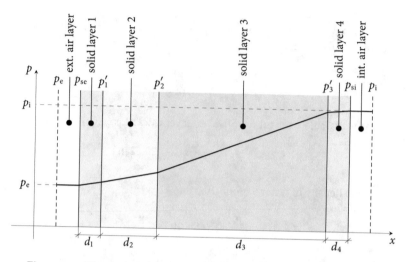

Figure 4.17: Water vapour pressure versus distance plot for a solid wall example. The thickness of air layers is informative.

acteristic water vapour pressure p'_n:

$$p'_n = p_e + \left(s_{d,se} + \sum_{j=1}^{n} s_{d,j} \right) \frac{p_i - p_e}{s_{d,tot}}. \tag{4.39}$$

Using this equation, we can avoid accumulating calculation errors.

Because standard ISO 13788 [44] strictly assumes that $s_{d,se} = s_{d,si} = 0$, equation (4.39) is simplified to

$$p'_n = p_e + \left(\sum_{j=1}^{n} s_{d,j} \right) \frac{p_i - p_e}{s_{d,tot}}. \tag{4.40}$$

Graphical method

By inspecting water vapour pressure versus distance plot in Fig. 4.17, we can see that the water vapour pressure function has different slopes for different layers. We can explain this fact by differentiating equation (4.38) as

$$\frac{\Delta p}{d} = \frac{g}{\delta_0} \mu \implies \frac{dp}{dx} = \frac{g}{\delta_0} \mu.$$

The slope of the function $p(x)$ is described by its first derivative and therefore proportional to the quotient of the density of water vapour flow rate by the water vapour permeability with respect to the vapour pressure of air g/δ_0 (the same for all layers) and proportional to the water vapour resistance factor μ (different for each layer).

We can, however, plot water vapour pressure versus equivalent air layer thickness (Fig. 4.18). In this case, from (4.38), we get

$$\frac{\Delta p}{s_d} = \frac{g}{\delta_0} \implies \frac{dp}{ds_d} = \frac{g}{\delta_0}.$$

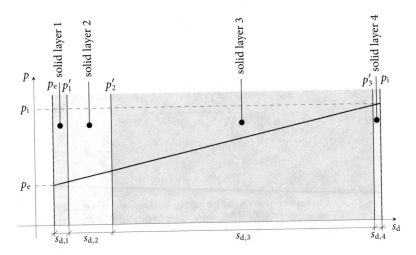

Figure 4.18: Water vapour pressure versus equivalent air layer thickness plot for a solid wall example. The function is linear, so we can use this plot to determine (read out) characteristic water vapour pressures graphically. Surface equivalent air layer thicknesses are so small that they cannot be seen on the plot.

We see that first derivative $p(s_d)$ and its slope are constant; that is, the function is linear. If we plot the surface equivalent air layer thickness of all layers (including air layers) on abscissa, we can simply connect external water vapour pressure p_e on one side with internal water vapour pressure p_i on the other side. Characteristic water vapour pressures then can be read out from the ordinate.

4.5.5 Multidimensional diffusion

If the water vapour does not flow perpendicularly to the surface, angle of incidence θ has to be taken into account (Fig. 4.19). Equation (4.3) is defined in terms of the surface perpendicular to water vapour flow of area A', so water vapour flow in the slanted case amounts to

$$q_m = A' g = A \cos \theta \, g.$$

We can define the water vapour flow rate as a vector

$$\vec{g} = g_x \vec{i} + g_y \vec{j} + g_z \vec{k}. \tag{4.41}$$

In this case, the water vapour flow rate can be rewritten in terms of a scalar product between those two vectors

$$q_m = \vec{A} \cdot \vec{g}, \tag{4.42}$$

where $\vec{A}$ is the surface vector (see Section 2.2.3 on page 36).

In one dimension, for thin layers, thickness and mass concentration of water vapour difference in (4.25) tend to zero, $d \to dx, \Delta v \to dv$, leading

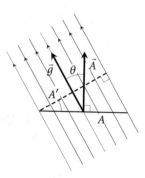

Figure 4.19: Nonperpendicular water vapour flow rate through a surface of area A. The scalar density of water vapour flow rate is defined in terms of the surface perpendicular to water vapour flow of area A', so angle of incidence θ has to be taken into account.

to differential version

$$g = -D\frac{\mathrm{d}v}{\mathrm{d}x},$$

(4.43)

where we have taken into account that when the mass concentration of water vapour increases in the $+x$ direction, water vapour flows in the opposite $-x$ direction and vice versa.

In more complex situations, however, the mass concentration of water vapour is a function of all three coordinates, $v(x, y, z)$, and the water vapour flows in all three directions. We must therefore write Fick's first law for each of dimensions

$$g_x = -\lambda\frac{\partial v}{\partial x}, \ g_y = -\lambda\frac{\partial v}{\partial y}, \ g_z = -\lambda\frac{\partial v}{\partial z}.$$

Using the density of water vapour flow rate vector (4.41) and *nabla operator* (2.19), we can transform Fick's first law into three-dimensional form

$$\vec{g} = -D\left(\frac{\partial v}{\partial x}\vec{i} + \frac{\partial v}{\partial y}\vec{j} + \frac{\partial v}{\partial z}\vec{k}\right),$$

$$\vec{g} = -D\vec{\nabla}v.$$

(4.44)

The density of water vapour flow rate is therefore the product of the water vapour diffusion coefficient and the negative *gradient* of the mass concentration of water vapour.

The gradient operator is elaborated on in Section 2.2.3 on page 36.

4.5.6 Dynamic diffusion

Until now, we have studied *steady diffusion*, that is, situations with time-independent mass concentration of water vapour that disregard the accumulation of water vapour in building components. Because the mass

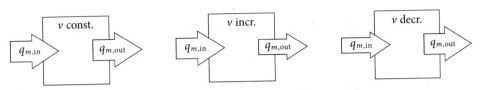

Figure 4.20: Difference between steady diffusion (left) and dynamic diffusion (middle, right). When the water vapour flow rate entering a building component differs from the one leaving, the mass concentration of water vapour changes.

concentration of water vapour in the building component is constant, the water vapour flow rate entering building component on one side is equal to the water vapour flow rate leaving the component on the other side (Fig. 4.20, left).

Now we will take a closer look at *dynamic diffusion*. Because the water vapour flow rate entering the building component on one side differs from the water vapour flow rate leaving the component on the other side, the mass concentration of water vapour of the building component must change (Fig. 4.20, middle and right).

In order to describe the mass concentration of water vapour change, we observe a small fragment of building component with dimensions $\Delta x \times \Delta y \times \Delta z$ (Fig. 4.21). In a dynamic situation, the water vapour flow rate that enters the fragment, $q_{m,in}$, is different from the water vapour flow rate that leaves it, $q_{m,out}$. The difference of water vapour flow rates corresponds to the net water vapour mass m transferred to/from the fragment in a given time period (4.1):

$$q_{m,net} = q_{m,in} - q_{m,out} = \frac{dm}{dt}.$$

Because the water vapour flow is not necessarily directed along one of the

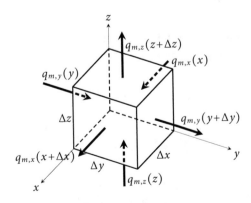

Figure 4.21: Small fragment of building component with dimensions $\Delta x \times \Delta y \times \Delta z$. In a dynamic situation, the water vapour flow rate that enters the fragment is different from the water vapour flow rate that leaves it, which leads to the change of mass concentration of water vapour.

coordinate axes, we have to decompose it into components:

$$\vec{q}_m = q_{m,x}\vec{i} + q_{m,y}\vec{j} + q_{m,z}\vec{k}.$$

As shown in Fig. 4.21, x, y and z components of water vapour flow rate $q_{m,x}$, $q_{m,y}$ and $q_{m,z}$ enter and leave the fragment along x, y and z axes, respectively. For simplicity, we will assume $q_{m,in} > q_{m,out}$ and consequently conclude that the net water vapour flow to the fragment increases its mass concentration of water vapour as in

$$\left[q_{m,x}(x) + q_{m,y}(y) + q_{m,z}(z)\right] - \left[q_{m,x}(x+\Delta x) + q_{m,y}(y+\Delta y) + q_{m,z}(z+\Delta z)\right] = \frac{\partial m}{\partial t}.$$

The water vapour mass can be written in terms of volume (4.8) as

$$m = v\,V = v\,\Delta x \Delta y \Delta z.$$

Because the water vapour flow rate is a smooth function, exiting water vapour flow rate components can be related to entering water vapour flow rate components using the Taylor series as

$$q_{m,x}(x+\Delta x) = q_{m,x}(x) + \frac{\Delta x}{1!}\frac{\partial q_{m,x}}{\partial x} + \frac{\Delta x^2}{2!}\frac{\partial^2 q_{m,x}}{\partial x^2} + \cdots \approx q_{m,x}(x) + \Delta x\frac{\partial q_{m,x}}{\partial x},$$

$$q_{m,y}(y+\Delta y) = q_{m,y}(y) + \frac{\Delta y}{1!}\frac{\partial q_{m,y}}{\partial y} + \frac{\Delta y^2}{2!}\frac{\partial^2 q_{m,y}}{\partial y^2} + \cdots \approx q_{m,y}(y) + \Delta y\frac{\partial q_{m,y}}{\partial y},$$

$$q_{m,z}(z+\Delta z) = q_{m,z}(z) + \frac{\Delta z}{1!}\frac{\partial q_{m,z}}{\partial z} + \frac{\Delta z^2}{2!}\frac{\partial^2 q_{m,z}}{\partial z^2} + \cdots \approx q_{m,z}(z) + \Delta z\frac{\partial q_{m,z}}{\partial z},$$

where we have neglected the higher order contributions. By using the preceding equations, we obtain

$$-\Delta x\frac{\partial q_{m,x}}{\partial x} - \Delta y\frac{\partial q_{m,y}}{\partial y} - \Delta z\frac{\partial q_{m,z}}{\partial z} = \Delta x \Delta y \Delta z\frac{\partial v}{\partial t}.$$

The density of water vapour flow rate is the water vapour flow rate divided by the corresponding cross-sectional area as

$$g_x = \frac{q_{m,x}}{\Delta y \Delta z}, \; g_y = \frac{q_{m,y}}{\Delta x \Delta z}, \; g_z = \frac{q_{m,z}}{\Delta x \Delta y},$$

leading to

$$\frac{\partial g_x}{\partial x} + \frac{\partial g_y}{\partial y} + \frac{\partial g_z}{\partial z} = -\frac{\partial v}{\partial t},$$

$$\vec{\nabla} \cdot \vec{g} = -\frac{\partial v}{\partial t}. \tag{4.45}$$

Here we used the definition of the density of water vapour flow rate as a vector (4.41) and nabla operator (2.19). This is the *mass continuity equation* in absence of mass sources, which states that the *divergence* of the density of water vapour flow rate is equal to the mass concentration of the water vapour change.

Joining the mass continuity equation (4.45) and the three-dimensional Fick's first law (4.44), we obtain *Fick's second law*

$$\vec{\nabla} \cdot (D\vec{\nabla}v) = \frac{\partial v}{\partial t}. \tag{4.46}$$

Fick's second law is a partial differential equation that describes the spatial and temporal variations of the mass density of water vapour v. It is also used to study water vapour mass transfer, where the equation is first solved to determine the mass density of water vapour, after which the density of water vapour flow rate is obtained using (4.44). Finally, the water vapour flow rate can be obtained by integrating the density of water vapour flow rate over surface A using the differential version of expression (4.42):

$$q_m = \int_A \vec{g} \cdot \mathrm{d}\vec{A}.$$

Note that, in general, the water vapour diffusion coefficient depends on spatial coordinates, as well as mass concentration of water vapour and temperature, $D = f(x, y, z, v, \theta)$, so solving differential equations for real problems can be extremely complicated. Usually, the solution is found by splitting building component into parts with the constant water vapour diffusion coefficient, in which case, equation (4.46) is simplified into the form

$$\frac{\partial v}{\partial t} = D\left(\frac{\partial^2 v}{\partial x^2} + \frac{\partial^2 v}{\partial y^2} + \frac{\partial^2 v}{\partial z^2}\right),$$

$$\frac{\partial v}{\partial t} = D\nabla^2 v, \tag{4.47}$$

where ∇^2 is the *Laplace operator* (2.25).

Because differential equations (4.46) and (4.47) include a time derivative, we also need the following to get the solution:

- *Boundary conditions*, value of the mass concentration of water vapour or the density of water vapour flow rate at the boundary of the observed system.

- *Initial conditions*, value of the mass concentration of water vapour within the observed system at the initial moment.

Practically all real problems are so complex that the solution to (4.46) and (4.47) must be found numerically.

4.6 Moisture transfer with condensation

The calculations in the previous sections were made under the assumption that water vapour pressure everywhere is smaller than water vapour pressure at saturation. However, this is often not the case, which leads to *interstitial condensation*.

This section discuss two different standardised methods that can be used to detect the possibility of water phase transitions and to determine the rate of condensation and rate of evaporation within the building component.

4.6.1 Glaser method

The simple and effective steady state assessment of condensation is obtained by the *Glaser method*.

According to standard ISO 13788 [44], calculations must be done for *monthly averages*. Input data include:

- the structure of the building component with the corresponding thermal conductivities and water vapour resistance factors,

- the monthly averages of external temperature and relative humidity, and

- the monthly averages of internal temperature and relative humidity. If data are unavailable, those could be calculated from the external temperature using recommendations in standard ISO 13788 [44] (Fig. 4.22).

The procedure involves the following steps:

Step 1: Calculate the characteristic temperatures (surface and interface temperatures) either by using (3.3), (3.5) and (3.6) or by using (3.7).

Step 2: Calculate the surface and interface water vapour pressures at saturation using (4.4).

Step 3: Determine the internal and external water vapour pressures from relative humidity using (4.7).

Step 4: Plot the water vapour pressure at saturation profile (surface and interface values) against equivalent air layer thickness $p_{sat}(s_d)$, calculated using (4.29).

Step 5: Plot the water vapour pressure profile against equivalent air layer thickness $p(s_d)$ by drawing the internal and external surface water vapour pressures and connecting them with a straight line (see Section 4.5.4 for the explanation).

Step 6: If condensation or evaporation sets in, modify the water vapour pressure profile, and calculate the rate of condensation or rate of evaporation.

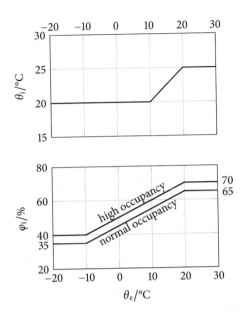

Figure 4.22: A simplified approach to determine internal boundary conditions (temperature and relative humidity) in dwellings and office buildings for the continental climate [44]. All concerned data are daily mean values.

Hereafter, we will consider three fundamental situations:

1. Water vapour diffusion without any interstitial condensation

2. Water vapour diffusion with one interstitial condensation

3. Water vapour diffusion with one interstitial evaporation

Water vapour diffusion without any interstitial condensation

In Fig. 4.23, water vapour pressure profile is completely below the profile at saturation $p < p_{sat} \implies \varphi < 100\,\%$. There is no interstitial condensation,

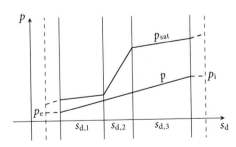

Figure 4.23: Water vapour diffusion without any interstitial condensation in the building component. The air layers plot is informative (dashed; equivalent air layer thickness is negligible).

and the density of water vapour flow rate through the building component can be calculated from (4.35) and (4.30) as

$$g = \delta_0 \frac{p_i - p_e}{s_{d,1} + s_{d,2} + s_{d,3}}.$$

Water vapour diffusion with the interstitial condensation

If we connect water vapour pressures on surface positions with a straight line in Fig. 4.24, it intersects with the water vapour pressure at saturation profile twice (dotted line). There is apparently a region for which the water vapour pressure is larger than the one at saturation, and condensation sets in at $p > p_{sat} \implies \varphi > 100\,\%$. But because the water vapour pressure cannot exceed that at saturation, we have to redraw the water vapour pressure profile as a series of lines that touch, but never go above, the water vapour pressure at the saturation profile (full line). We can now calculate the density of water vapour flow rate that enters the interface with condensation as

$$g_{in} = \delta_0 \frac{p_i - p_c}{s_{d,2} + s_{d,3}}$$

and the density of water vapour flow rate that leaves the interface with condensation as

$$g_{out} = \delta_0 \frac{p_c - p_e}{s_{d,1}},$$

where p_c is the water vapour pressure at saturation for the interface with condensation. Because the slope representing the former (g_{in}) is larger than the slope representing the latter (g_{out}), the water vapour flow that enters the interface exceeds the flow that leaves the interface and liquid water at the interface must accumulate. The rate of condensation is equal to the difference of the density of the water vapour flow rates:

$$g_c = g_{in} - g_{out} = \delta_0 \left(\frac{p_i - p_c}{s_{d,2} + s_{d,3}} - \frac{p_c - p_e}{s_{d,1}} \right).$$

Finally, by taking one month as the condensation time, we can calculate the amount of condensed water at the interface of the building component

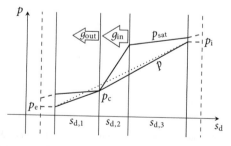

Figure 4.24: Water vapour diffusion with interstitial condensation at one interface of the building component. The air layers plot is informative (dashed; equivalent air layer thickness is negligible).

using (4.2) and (4.3):

$$\Delta \rho_A = \frac{m}{A} = g_c\, t.$$

Quantity ρ_A represents the water surface density (mass of water per area of building component).

Water vapour diffusion with the interstitial evaporation

As shown in Fig. 4.25, when there is liquid water already accumulated in an interface, the vapour pressure is equal to that at saturation, even if the straight line that connects the surface water vapour pressures (dotted line) does not intersect with the water vapour pressure at saturation profile. We therefore have to redraw the water vapour pressure profile as a series of lines that touch the water vapour pressure at the saturation profile at the interfaces with accumulated water. When all of the accumulated water evaporates, the Glaser diagram returns to the state shown in Fig. 4.23.

We can now calculate the density of water vapour flow rate that enters the interface with evaporation as

$$g_{in} = \delta_0 \frac{p_i - p_c}{s_{d,2} + s_{d,3}}$$

and the density of water vapour flow rate that leaves interface with evaporation as

$$g_{out} = \delta_0 \frac{p_c - p_e}{s_{d,1}},$$

where p_c is the water vapour pressure at saturation for the interface with evaporation. Because the slope representing the former (g_{in}) is smaller than the slope representing the latter (g_{out}), the water vapour flow that leaves the interface exceeds the flow that enters the interface, and the liquid water amount at the interface must decrease. The rate of evaporation is equal to the difference of the density of water vapour flow rates:

$$g_{ev} = g_{out} - g_{in} = \delta_0 \left(\frac{p_c - p_e}{s_{d,1}} - \frac{p_i - p_c}{s_{d,2} + s_{d,3}} \right).$$

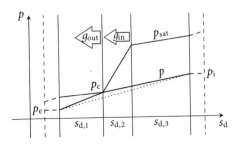

Figure 4.25: Water vapour diffusion with evaporation at one interface of the building component with one direction flow. The air layers plot is informative (dashed; equivalent air layer thickness is negligible).

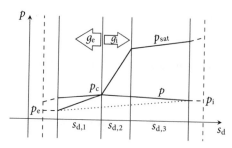

Figure 4.26: Water vapour diffusion with evaporation at one interface of the building component with two directions flow. The air layers plot is informative (dashed; equivalent air layer thickness is negligible).

It is also possible that the evaporated water leaves the interface in both directions, as shown in Fig. 4.26.

We can now calculate the density of water vapour flow rate that leaves the interface towards interior as

$$g_i = \delta_0 \frac{p_c - p_i}{s_{d,2} + s_{d,3}}$$

and the density of water vapour flow rate that leaves the interface towards exterior as

$$g_e = \delta_0 \frac{p_c - p_e}{s_{d,1}},$$

where p_c is the water vapour pressure at saturation for the interface with evaporation. The rate of evaporation is equal to the sum of the density of water vapour flow rates:

$$g_{ev} = g_e + g_i = \delta_0 \left(\frac{p_c - p_e}{s_{d,1}} + \frac{p_c - p_i}{s_{d,2} + s_{d,3}} \right).$$

Finally, by taking one month as the evaporation time, we can calculate the amount of evaporated water at the interface of the building component using (4.2) and (4.3):

$$\Delta \rho_A = \frac{m}{A} = g_{ev}\, t.$$

As pointed out previously, the procedure must be repeated for every month of the year (Fig. 4.27). The amounts of condensed and evaporated water for individual months provide the *total water surface density* through the whole year by adding condensed water amounts and subtracting evaporated water amounts month by month (Table 4.2).

Note that standard ISO 13788 [44] demonstrates more examples for interstitial condensation and evaporation, including condensation and evaporation on several interfaces. The standard also recommends that items with thermal resistance greater than 0.25 m² K/W are subdivided into a number of notional layers, each with thermal resistance not exceeding 0.25 m² K/W.

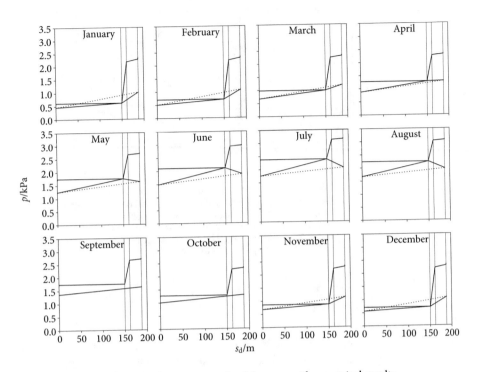

Figure 4.27: Glaser diagrams for every month of the year, with numerical results in Table 4.2.

month	$\Delta\rho_A \left/ \frac{g}{m^2}\right.$	$\rho_A \left/ \frac{g}{m^2}\right.$
November	3.23	3.23
December	4.52	7.75
January	5.10	12.85
February	4.34	17.19
March	1.76	18.95
April	−1.20	17.75
May	−3.30	14.45
June	−5.22	9.23
July	−6.48	2.75
August	−2.75	0.00
September	0.00	0.00
October	0.00	0.00

Table 4.2: Water surface density through the year from the Glaser diagrams in Fig. 4.27.

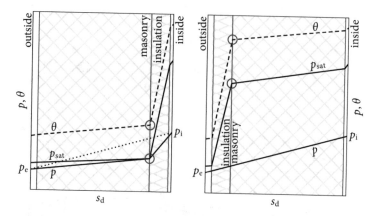

Figure 4.28: The Glaser diagram and temperature for insulation on the inner side (left) and the outer side (right) for a typical building component with masonry. By putting insulation on the outer side of the building component, the temperature as well as water vapour pressure at saturation are substantially enlarged on the interface between the insulation and masonry, which prevents condensation.

Position of thermal insulator in the building component

In continental climates, the probability of interstitial condensation is the largest in winter months, when the internal temperature and water vapour pressure are larger than the external ones. The probability of condensation can be significantly reduced by ensuring that the layer with the smallest thermal conductivity (thermal insulator) is put on the outer and not inner side of the building component. We will show the reasoning for this based on a typical building component with masonry.

> **Info box**
>
> By putting the thermal insulator on the outer side of the masonry, we can increase the temperature in the building component and prevent condensation.

The Glaser diagrams and temperatures for a building components with masonry and insulation layers for typical winter continental conditions are presented in Fig. 4.28. When insulation is on the inner side (left), condensation appears on the interface between the insulation and masonry. By putting insulation on the outer side of the building component (right), condensation is prevented. Temperature as well as water vapour pressure at saturation are substantially enlarged on the interface between the insulation and masonry.

4.6.2 Dynamic moisture and heat transfer

As we pointed in Section 2.2.4 on page 38 and in Section 4.5.6, *heat transfer and water vapour transfer are coupled,* as both thermal conductivity and the diffusion coefficient are dependent on temperature and moisture content. Furthermore, in Section 4.6.1, we have already studied additional complications arising due to water vapour *condensation* or liquid water *evaporation* within the building component. Finally, water from interstitial condensation as well as surface water (precipitation, sorption, condensation) can move by means of *capillary action* across the component. To simulate real problems, all of these concerns have to be taken into account.

Standard EN 15026 [17] elaborates on the numerical simulation of one-dimensional dynamic moisture and heat transfer. We will present some highlights of this method in this section.

Input data include the structure of the building component with corresponding thermal conductivities, water vapour resistance factors, sorption isotherms and liquid water conductivities. Further data requirements include environmental variables, external and internal temperatures, external and internal relative humidities, external radiation data, precipitation, and wind speed and direction.

Within the building element, we observe variation of temperature θ, as well as variation of moisture quantities, mass concentration of water w, relative humidity φ or water vapour pressure p. The latter three are connected by sorption isotherm $w(\varphi)$ (Fig. 4.9 on page 132), where hysteresis is to be neglected, and by definition of relative humidity (4.7) as

$$\varphi = \frac{p}{p_{\text{sat}}(\theta)},$$

where the water vapour pressure at saturation p_{sat} is obtained from the temperature.

Heat transport by thermal conduction

We first assume that thermal conductivity does not depend on temperature but only on mass concentration of water $\lambda(w)$, so the sensible density of heat flow rate is a one-dimensional version of (2.20):

$$q_{\text{sens}} = -\lambda(w)\frac{d\theta}{dx}. \tag{4.48}$$

Moisture transport by vapour diffusion

We have already taken into account the strong temperature dependence of the diffusion coefficient by introducing water vapour permeability δ_0 (4.31); however, its dependence on a mass concentration of water vapour is still contained in vapour resistance factor $\mu(\varphi)$. The density of water vapour flow rate is then described by a differential version of (4.30):

$$g = -\frac{\delta_0}{\mu(\varphi)}\frac{dp}{dx}. \tag{4.49}$$

Heat transport by vapour diffusion

Next, we take into account the condensation and the evaporation, which change the energy of the material fragment. As we have elaborated on in Section 4.6.1, in quasi-stationary cases, the net inflow of water vapour is caused by condensation to liquid water, whereas the net outflow of water vapour is caused by evaporation of the liquid water. The former clearly increases the energy of the fragment, whereas the latter decreases the energy. By dividing expression (1.30) with time and area, for the latent density of heat flow rate, we get

$$q_{lat} = h_v\, g, \tag{4.50}$$

where h_v is the specific enthalpy of vaporisation and g is rate of condensation/evaporation.

Moisture transport by liquid water transport

We also take into account water transfer due to capillary suction. We describe this transfer by the density of liquid water flow rate as

$$g_w = K(p_{suc})\frac{dp_{suc}}{dx}, \tag{4.51}$$

where K (s) is *liquid water conductivity* and p_{suc} (Pa) is *suction pressure*, that is the difference of ambient atmospheric pressure and water pressure inside the pores. Suction pressure can be calculated using the Kelvin equation (4.22):

$$p_{suc} = -\rho R_v T \ln \varphi.$$

Energy storage

The energy storage is described by a one-dimensional version of heat continuity equation (2.21) as

$$\frac{d(q_{sens} + q_{lat})}{dx} = -\left(\rho_w c_w + \rho_m c_m\right)\frac{dT}{dt}, \tag{4.52}$$

where c_w and c_m are specific heat capacities of water and dry material, respectively, and ρ_w and ρ_m are densities of water and dry material, respectively.

Moisture storage

Moisture storage is described by a one-dimensional version of mass continuity equation (4.45) as

$$\frac{d(g + g_w)}{dx} = -\frac{dw}{dt}. \tag{4.53}$$

Surface and interface conditions

The standard prescribes that surface conditions are calculated with exact methods. Surface resistances are calculated using expressions (3.1), (2.47) and values from Table 2.3 on page 51, also taking into account sol-air temperature (3.1.5). On the other hand, surface equivalent air layer thicknesses are to be taken from Table 4.1 on page 140. Additional liquid water flow resistance on the interface between two materials can also be included.

Finally, the uptake of rain is also taken into account, with consideration of the maximum possible density of liquid water flow due to precipitation.

4.7 Requirements concerning building components

In order to avoid causes for excessive moisture, as categorised in Section 4.1, building components must fulfil these conditions (Fig. 4.2 on page 119):

A Liquid water intrusion must be completely prevented.

B Relative humidity at the internal surface should be below the limit.

C Build-up of moisture due to water diffusion should be below the limit.

D Building materials used for construction must be dry enough.

Liquid water intrusion must be completely prevented

For new buildings, parts of the building that are in contact with ground water (foundations, basement walls, green roofs) must be insulated using waterproofing materials. Gravel below the basement slab helps prevent capillary transport, as capillary action is negligible due to large voids (Section 4.4.2). Another important measure is a water draining system, which transfers liquid water away from the building.

For existing houses with capillary moisture problems (Section 4.4.2) several mitigation measures are possible:

- Mechanical barriers: We interrupt capillaries by cutting the building walls horizontally near the basement slab and inserting waterproofing material.

- Chemical barriers: Materials are treated by chemicals that change the properties of pores, for example, increasing the contact angle (Section 4.4.2).

The presence of precipitation is addressed differently. Typical mitigations are:

- Installing water vapour permeable membranes. If façades were hermetically sealed for water transfer, including water vapour, condensation would appear on the internal side of the water barrier (Section 4.6.1). Water vapour permeable membranes prevent the transfer of liquid water and still allow water vapour to penetrate.

- Using façade materials of low sorptivity to prevent capillary transport to interior.

- Eaves help keeping higher parts of the façade dry.

Relative humidity at the internal surface should be below the limit

Temporary condensation, that is, the internal surface relative humidity of 100 %, is allowed only for those building components that do not transfer water because of their low porosity or special protective layers (glass, window frame). On the other hand, in order to avoid mould growth, standard ISO 13788 [44] prescribes that the monthly internal surface relative humidity for other building components should not exceed the critical relative humidity of 80 %. Due to hygroscopy (Section 4.4.1), the building component contains water vapour even if the relative humidity is $\varphi < 100$ %, so mould growth already sets in at lower relative humidities. Note that this condition is related to the minimum temperature factor of the internal surface (Section 4.5.3).

Both problems can be solved using the following methods:

- Increase the internal surface temperature. This is achieved by increasing the total thermal resistance and consequently the temperature factor of the internal surface of the building component, or by heating external walls.

Info box

Due to the many water vapour sources in the building, the water vapour has to be constantly eliminated, usually by ventilation.

- Decrease the relative humidity of the internal air. Evaporation from liquid water sources and living beings continuously increases the water vapour amount (Table 4.3). This problem can be mitigated by ventilation (Section 3.3.4 on page 108), which replaces humid internal air with drier external air, or by air conditioning, which extracts water vapour from the internal air.

Moisture Source	Contribution
human insensible loss	1.2 L/d
- *of that*, perspiration	0.4 L/d
- *of that*, respiration	0.4 L/d
cooking	3 L/d
showering	0.5 L/d
dish washing	0.5 L/d
clothes washing	0.5 L/d
drying new construction materials	5 L/d

Table 4.3: Moisture sources in the building with typical contributions [79].

Build-up of moisture due to water diffusion should be below the limit

It is recommended to completely prevent interstitial condensation due to water diffusion (Section 4.6.1). However, condensation is allowed under the following conditions:

- The water quantity of the building component should not exceed the limit value, usually defined in terms of maximum water mass ratio u_{max}. Limit values are usually set to the value at which capillary water transport within the building component sets in.

- The building component must be dry (without liquid water) for at least one month per year [44].

The problem can be addressed in different ways, depending on the environmental conditions. For continental climates, the most effective mitigations are

- Putting thermal insulator on the outer side of the masonry, as explained in Section 4.6.1, and

- Installing water vapour barriers. Water vapour barriers prevent the transfer of water vapour into the building component. Because the internal mass concentration of water vapour is larger than the external concentration (Example 4.3), the water vapour barrier has to be positioned on the inner side of the building component.

- Installing water vapour permeable membranes. Water vapour permeable membranes prevent the transfer of liquid water and still allow water vapour to penetrate. They are positioned on the outer side of the building component, protecting it from the ingress of liquid water, such as precipitation, while still facilitating the transfer of water vapour out of the building component.

Building materials used for construction must be dry enough

Building materials that are being mounted during the construction, such as timber, should be appropriately dried beforehand. On the other hand, materials that are cast at the construction site, such as concrete, should be left to dry appropriately before the next phases of construction proceed.

4.8 Comfort conditions

The perceived comfort within the internal environment is affected by a combination of several thermodynamic variables, including air humidity. The perception is person dependent, but studies have identified key variables and the values that provide comfort for the greatest possible number of people.

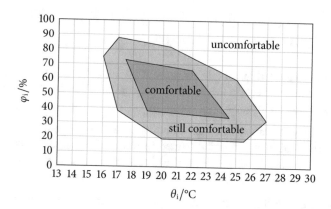

Figure 4.29: Comfort in relation to internal air temperature and internal relative humidity [68].

The following are the most important variables:

- *internal air temperature* θ_i and

- *internal relative humidity* φ_i.

The conditions for these two variables are usually interconnected (Fig. 4.29).

Standard ISO 7730 [29] quantifies the thermal comfort more precisely by defining the predicted mean vote (PMV), which should be as close as possible to zero. The PMV value is determined using a complicated expression, which apart from θ_i and φ_i, also takes into consideration other important variables:

- *Air velocity* is important because it influences the convective heat flow rate between human skin and internal air (see Section 2.3.1 on page 49). For example, a high velocity of cold internal air causes faster skin cooling and an unpleasant chilly effect.

- *Radiant temperature*, that is, the temperature of internal surfaces, is important because it influences the radiative heat flow rate between human skin and internal surfaces (see Section 2.4.3 on page 64). For example, even if internal air is warm, the low surface temperature of external building elements θ_{si} causes faster skin cooling and an unpleasant chilly effect.

- *Metabolic rate* in W/m^2 is important because more intense activity increases internal body temperature.

- *Clothing insulation* in $m^2\,K/W$ is important because better insulation reduces the conductive heat flow between the human skin and the outer clothing surface (see Section 2.2.1 on page 32).

Finally, *thermal effusivity* b is important when human skin comes in contact with the building element. This causes conductive heat flow, which is dependent on the building element thermal effusivity value (see Section 2.2.5 on page 42).

Problems

4.1 Humid air of volume $1.0\,m^3$, temperature $20\,°C$ and pressure $1.0\,bar$ contains $9.0\,g$ of water vapour. Air is compressed isothermally. At which pressure does the air become saturated with water vapour? Take the molar mass of water to be $18\,g/mol$. ($1.9\,bar$)

4.2 Using the psychrometric chart, determine the temperature of the glass at which condensation occurs, for humid room air at temperature $30\,°C$ and relative humidity $40\,\%$. ($15\,°C$)

4.3 An air dryer, which is working on the principle of removing heat, is used to eliminate the excess moisture in an $8.0\,m \times 6.0\,m \times 3.0\,m$ room. The initial relative humidity is $70\,\%$, the initial and final air temperatures are $30\,°C$ and $11\,°C$, respectively. Using the psychrometric chart, determine the mass of eliminated water. ($1.8\,kg$)

4.4 In the air conditioner, air of mass $100\,kg$ at temperature $30\,°C$ and relative humidity $60\,\%$ is first cooled and then warmed, so the final temperature is $20\,°C$ and relative humidity is $65\,\%$. Using the psychrometric chart, calculate the removed heat, the added heat and the mass of extracted water. ($3.35\,MJ$, $0.69\,MJ$; $660\,g$)

4.5 Air of mass $50\,kg$, which includes water vapour of mass $700\,g$, at temperature $20\,°C$ is mixed adiabatically with air of mass $33\,kg$ at temperature $25\,°C$ and relative humidity $30\,\%$. Using the psychrometric chart, determine the final temperature and relative humidity. ($22\,°C$, $65\,\%$)

4.6 Consider damp timber of density $450\,kg/m^3$ and mass ratio of water to dry matter of $20\,\%$. Calculate the mass concentration of water of the damp timber and density of the completely dry timber. ($75\,kg/m^3$, $380\,kg/m^3$)

4.7 The thermal transmittance of a vertical building component is $2.0\,W/(m^2\,K)$. Take the internal relative humidity to be $65\,\%$ and the internal and external temperature to be $20\,°C$ and $-5\,°C$, respectively. Calculate the relative humidity on the internal surface of the building component. ($98\,\%$)

4.8 A vertical building component separates the internal environment at temperature $20\,°C$ and relative humidity $70\,\%$ from the external environment at temperature $0\,°C$. Calculate the minimal total thermal resistance of the building element that prevents mould growth. Mould growth occurs for relative humidities above $80\,\%$. ($1.21\,m^2\,K/W$)

4.9 The temperature factor of the internal surface for the thermal bridge is 0.71. Assuming that the internal temperature is 21 °C and internal relative humidity is 55 %, calculate the external temperature for which mould growth occurs. Mould growth occurs for relative humidities above 80 %. (0.4 °C)

4.10 For all three situations (|stone|concrete|, |stone|concrete|EPS|, |stone|EPS|concrete|) from problem 3.3 in Chapter 3 calculate the total equivalent layer thickness, characteristic water vapour pressures at saturation and characteristic water vapour pressures, all under the assumption that there is no condensation. Internal and external relative humidities are both 65 %, the external and internal temperatures are −5 °C and 20 °C, respectively. Water vapour resistance factors are 120 for concrete, 200 for façade stone and 60 for expanded polystyrene (EPS). Plot the water vapour pressure and water vapour pressure at saturation as a function of the layer thicknesses and as a function of the equivalent air layer thicknesses. In which situation does condensation appear and where? (28, 37, 37; 401 Pa, 501 Pa, 530 Pa, 1352 Pa, 2337 Pa; 401 Pa, 409 Pa, 410 Pa, 450 Pa, 2238 Pa, 2337 Pa; 401 Pa, 409 Pa, 410 Pa, 2093 Pa, 2238 Pa, 2337 Pa; 261 Pa, 441 Pa, 1519 Pa; 261 Pa, 397 Pa, 1213 Pa, 1519 Pa; 261 Pa, 397 Pa, 703 Pa, 1519 Pa)

4.11 Vertical walls of the thermal envelope consists of (from outside to inside)

- fibre cement board of thickness 2.0 cm, thermal conductivity 1.5 W/(m K) and water vapour resistance factor 50;
- concrete of thickness 20.0 cm, thermal conductivity 1.0 W/(m K) and water vapour resistance factor 120; and
- expanded polystyrene of thickness 8.0 cm, thermal conductivity 0.035 W/(m K) and water vapour resistance factor 60.

The average monthly internal and external temperatures are 20 °C and 0 °C, respectively, the average monthly internal and external relative humidities are 45 % and 80 %, respectively. Using the Glaser method, determine whether condensation appears, and, if so, calculate the mass of water vapour condensed in one month with 30 days. (yes, 34 g/m^2)

4.12 The vertical walls of the thermal envelope consists of

- brick of thickness 10 cm, thermal conductivity 0.16 W/(m K) and water vapour resistance factor 16;
- mineral wool of thickness 10 cm, thermal conductivity 0.035 W/(m K) and water vapour resistance factor 1; and
- brick of thickness 10 cm, thermal conductivity 0.16 W/(m K) and water vapour resistance factor 16.

The average monthly internal and external temperatures are 20 °C and 5 °C, respectively, and the average monthly internal and external relative humidities are 50 % and 80 %, respectively. If condensed water exists in the most sensitive interstice, determine the mass of the evaporated water in one month with 30 days using the Glaser method. ($62 \, \text{g/m}^2$)

4.13 In a room of rectangular floor plan with dimensions 4.0 m × 5.0 m and height 2.5 m, there is dry air of density $1.20 \, \text{kg/m}^3$ at temperature 21 °C. In the room we place a pot with 3.0 L of liquid water at temperature 41 °C. What is the mass of evaporated water at the moment when temperatures of air and liquid water in the room drop to 16 °C? What is the relative humidity then? The specific heat capacity of dry air is $1.0 \, \text{kJ/(kg K)}$, the specific heat capacity of liquid water is $4.2 \, \text{kJ/(kg K)}$ and the specific heat of evaporation of water at 16 °C is $2460 \, \text{kJ/kg}$. The molar mass of water is 18 g/mol. (0.25 kg, 37 %)

5 Basics of waves

5.1 Disturbance and pulse

We usually associate the concept of 'wave' to the ridge and swell on the surface of a body of water. When water is disturbed, ridges and swells travel on its surface, but the bulk of the water does not move (Fig. 5.1).

Waves in the physical sense are any propagation of disturbance through space without medium transport; that is, particles of medium that transfer disturbance are not displaced, but rather oscillate around a fixed (equilibrium) position. Waves are extremely important because they facilitate energy propagation.

Info box

Waves are important because they transfer energy.

In terms of the physical mechanism, there are two types of waves:

1. *Mechanical waves* propagate through a medium by oscillation of material particles. The most prominent example of a mechanical wave is sound.

2. *Electromagnetic waves* do not require a medium for propagation. They consist of oscillations of electrical and magnetic fields. The most prominent examples are radio waves, infrared radiation, visible light and ultraviolet radiation.

On the other hand, in terms of particle/field oscillations, there are two main types of waves:

Figure 5.1: Water waves are disturbances that are travelling on the water surface from the point of the excitation. Note that for point excitation (droplet), waves form circular shapes.

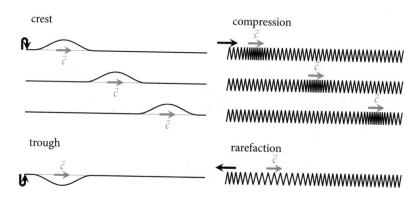

Figure 5.2: Transversal pulses (left) and longitudinal pulses (right).

A. *Transversal waves* consist of particle/field oscillations perpendicular to the direction of the wave propagation.

B. *Longitudinal waves* consist of particle/field oscillations parallel to the direction of the wave propagation.

Other wave types exhibit a combination of displacements parallel and perpendicular to the direction of wave propagation. These include water surface waves and earthquake waves.

We can demonstrate a transversal wave on a string under tension. If we strike one side of the string upward once, a *pulse* in the form of the crest is formed and travels along the string (Fig. 5.2, left). Note that despite pulse propagation, the string itself is not displaced. We can also create a pulse in the form of a trough by striking the string downward.

We can demonstrate the longitudinal wave on a stretched spring. If we strike one side of the spring to the right once, a *pulse* in the form of a spring compression is formed and travels along the spring (Fig. 5.2, right). Note that despite the pulse propagation, the spring undergoes only a small net movement. If we strike the spring to the left, a pulse in the form of the spring rarefaction is created.

In order to obtain a mathematical description of pulse propagation, let's observe a transversal pulse travelling on a string to the right along the z-axis as shown in Fig. 5.3. The left side of the figure shows the pulse at initial time $t = 0$, and the right side shows the pulse at time t. The assumption is that the pulse shape does not change.

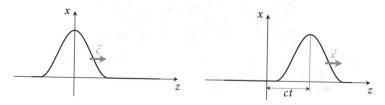

Figure 5.3: Pulse in the initial moment (left) and after time t (right).

At the initial moment, the pulse is represented by a certain mathematical function

$$x(z,0) = f(z),$$

where $x(z,0)$ is *displacement* from the equilibrium position, at location z in the initial moment.

If the pulse speed is equal to c, after time t, the pulse has moved to the right by distance ct, therefore, the pulse is represented by

$$x(z,t) = f(z - ct), \tag{5.1}$$

where $x(z,t)$ is *displacement* from the equilibrium position at location z in moment t.

Similarly, if the pulse travels to the left, the pulse is represented by

$$x(z,t) = f(z + ct). \tag{5.2}$$

The function $x(z,t)$ is usually called the *wave function* and depends on both the spatial z and temporal t variables. Note that for a longitudinal pulse, displacement x and position z have the same direction, whereas for the transversal pulse, displacement x is perpendicular to position z.

5.2 Travelling waves

By definition a *travelling wave* is a *periodic* disturbance propagating through space. The most important type of wave function is the travelling harmonic wave, that is, the wave described in terms of harmonic sine and cosine functions. It turns out that any nonharmonic wave can be expanded into a series of harmonic waves using Fourier analysis, so knowledge of harmonic waves is sufficient for describing any travelling wave.

Figure 5.4 presents transversal (top) and longitudinal (bottom) harmonic waves. Transversal waves can be created by harmonically pushing a string under tension up and down, resulting in alternating crests and troughs. Longitudinal waves can be created by harmonically pushing a stretched spring left and right, resulting in alternating compressions and rarefactions. We define *wavelength* λ (m) as

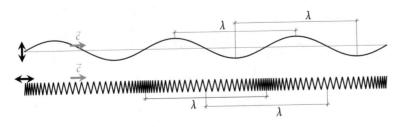

Figure 5.4: A travelling transversal (top) and longitudinal (bottom) harmonic waves. Wavelength λ is the distance between two crests or two troughs for transversal waves and between two compressions or two rarefactions for longitudinal waves.

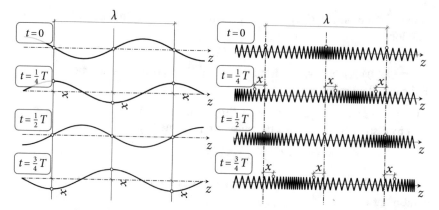

Figure 5.5: Movements of three selected points for travelling transversal (left) and longitudinal (right) harmonic wave, at four characteristic moments. Points oscillate around an equilibrium position denoted by the dash-dotted line with the same period but a different phase. Note that the left and right points are in phase (synchronised), whereas the middle point is in counter-phase to those points.

- the distance between two adjacent crests or two adjacent troughs for transversal waves and

- the distance between two adjacent compressions or two adjacent rarefactions for longitudinal waves.

More generally, the wavelength is the distance between two adjacent points that oscillate *in phase*, meaning that their movements are synchronised.

In order to describe waves mathematically, we will use standard kinematic expressions for oscillation.

We start by inspecting the movement of an arbitrary point of string under tension (Fig. 5.5, left) or an arbitrary point of stretched spring (Fig. 5.5, right). As shown, the point oscillates around a equilibrium position denoted by the dash-dotted line. The time required for the point to make one cycle of movement is called *period* T (s). Period is related to *frequency* f (Hz) and *angular frequency* ω (1/s) as

$$T = \frac{1}{f} = \frac{2\pi}{\omega}. \tag{5.3}$$

Note that oscillations of all points have the same period but different oscillation phases.

As with the pulse, the point displacement from the equilibrium position is a function of location z and time t, $x(z, t)$. First, we describe the point oscillation at location $z = 0$ as

$$x(0, t) = x_0 \sin(\omega t + \varphi), \tag{5.4}$$

where x_0 is the *amplitude* of oscillation and φ is its phase constant. Amplitude represents the maximum point displacement.

At moment $t = 0$, displacement of that point is $x_0 \sin \varphi$. Because the wave moves to the right with speed c, another point at location z will have the same displacement at time $t = z/c$:

$$x(z, t) = x_0 \sin \left[\omega \left(t - \frac{z}{c} \right) + \varphi \right].$$ (5.5)

Defining *angular wavenumber* k (1/m) as

$$k = \frac{\omega}{c},$$ (5.6)

choosing the starting moment so that $\varphi = \pi$ and using trigonometric identity $\sin(\alpha + \pi) = \sin(-\alpha)$, we finally get

$$x = x_0 \sin(kz - \omega t).$$ (5.7)

This is the *travelling harmonic wave* function in the right direction. It can be shown that the travelling harmonic wave function in the left direction is

$$x = x_0 \sin(kz + \omega t).$$ (5.8)

Note the similarities between the wave functions for the travelling wave, (5.7) and (5.8), and those for travelling pulse, (5.1) and (5.2).

As pointed out previously and shown in Fig. 5.5, any two points distanced by wavelength λ are in phase (synchronised). Using identity $\sin(\alpha + 2\pi) = \sin(\alpha)$, we conclude that the product of the wavelength and angular wavenumber must be

$$k\lambda = 2\pi.$$ (5.9)

Using (5.6), (5.3) and (5.9), we can write the wave speed as the product of the frequency and wavenumber:

$$c = f\lambda.$$ (5.10)

5.3 Wave equation

In this section, we will derive the wave equation for a string under initial tension F_0. Let's inspect the movement of a small fragment of a string with mass Δm. This is a case of transversal waves, so the movement of the fragment is exclusively perpendicular to the undisturbed string as shown in Fig. 5.6. Because the string fragment does not move parallel to the undisturbed string (z-axis), according to Newton's second law, the net force on the string element in this direction must be zero as in

$$F_2 \cos \alpha_2 - F_1 \cos \alpha_1 = 0.$$

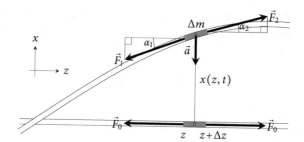

Figure 5.6: Small fragment of the string under tension in equilibrium position and displaced due to wave propagation.

The cosine component of the tension is constant for the whole string and therefore is equal to the tension of the undisturbed string F_0 as in

$$F_1 \cos \alpha_1 = F_2 \cos \alpha_2 = F_0.$$

On the other hand, the displaced string fragment accelerates perpendicular to the undisturbed string (x-axis), so according to Newton's second law, the net force in this direction is proportional to string fragment acceleration a as in

$$F_2 \sin \alpha_2 - F_1 \sin \alpha_1 = \Delta m\, a.$$

Combining both equations, we get

$$F_0(\tan \alpha_2 - \tan \alpha_1) = \Delta m\, a.$$

Because the disturbed string is described by function $x(z, t)$, $\tan \alpha$ is equal to the spatial partial derivative of that function, $\partial x/\partial z$:

$$F_0 \left[\frac{\partial x}{\partial z}\Big|_{z+\Delta z} - \frac{\partial x}{\partial z}\Big|_{z} \right] = \Delta m\, a. \tag{5.11}$$

Both displacement $x(z, t)$ and its derivatives are smooth functions. We can define new function f as

$$\frac{\partial x}{\partial z} = f, \quad \frac{\partial x}{\partial z}\Big|_{z} = f(z), \quad \frac{\partial x}{\partial z}\Big|_{z+\Delta z} = f(z+\Delta z).$$

Because f is a smooth function of position z, we can relate the function values at two close positions using the Taylor series

$$f(z+\Delta z) = f(z) + \frac{\Delta z}{1!}\frac{\partial f}{\partial z} + \frac{\Delta z^2}{2!}\frac{\partial^2 f}{\partial z^2} + \cdots \approx f(z) + \Delta z\frac{\partial f}{\partial z}$$

$$\implies f(z+\Delta z) - f(z) = \Delta z\frac{\partial f}{\partial z},$$

or in terms of the spatial partial derivative as

$$\frac{\partial x}{\partial z}\Big|_{z+\Delta z} - \frac{\partial x}{\partial z}\Big|_{z} = \Delta z\frac{\partial^2 x}{\partial z^2}.$$

Inserting this into (5.11), we get

$$F_0 \Delta z \frac{\partial^2 x}{\partial z^2} = \Delta m\, a.$$

Noting that acceleration a is the second derivative of displacement

$$a = \frac{\partial^2 x}{\partial t^2}, \tag{5.12}$$

and defining linear density ρ_l (kg/m) as

$$\rho_l = \frac{\Delta m}{\Delta z},$$

we finally get

$$\frac{\partial^2 x}{\partial t^2} = \frac{F_0}{\rho_l} \frac{\partial^2 x}{\partial z^2}. \tag{5.13}$$

We can easily verify that the wave functions for the travelling harmonic wave, (5.7) and (5.8), are solutions of this equation by inserting them into (5.13). Because

$$\frac{\partial^2}{\partial t^2} \left[\sin(kz \pm \omega t) \right] = -\omega^2 \sin(kz \pm \omega t),$$

$$\frac{\partial^2}{\partial z^2} \left[\sin(kz \pm \omega t) \right] = -k^2 \sin(kz \pm \omega t),$$

we get

$$\omega^2 = \frac{F_0}{\rho_l} k^2. \tag{5.14}$$

Using (5.6), we obtain the expression for the wave speed in the string under tension:

$$c^2 = \frac{F_0}{\rho_l}. \tag{5.15}$$

Inserting the obtained expression back into (5.13), we get the basic form of the *wave equation*:

$$\frac{\partial^2 x}{\partial t^2} = c^2 \frac{\partial^2 x}{\partial z^2}. \tag{5.16}$$

Wave functions for travelling pulse, (5.1) and (5.2), are also solutions of the wave equation.

5.4 Pulse interactions

So far, we have been concerned with travelling pulses and waves, that is, pulses and waves propagating through an ideal (infinite) uniform medium without dissipation or interaction with other entities. However, when pulses or waves encounter a change in the medium or another pulse, the interaction leads to new phenomena, that is, reflection, transmission and superposition.

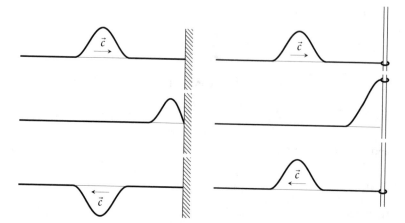

Figure 5.7: Total reflection for a fixed string end (left) and a free string end (right). In both cases, the amplitude of the pulse does not change; however, in the former case, displacement is inverted.

5.4.1 Reflection

First, we consider the interaction of a pulse with a *perfect boundary*. This type of boundary corresponds to the end of a string of finite dimensions. There is no way a pulse can propagate beyond the end of the string, so it bounces back and travels in the opposite direction. The implicated process is called *reflection*.

There are two types of perfect boundaries, depending on the string termination method:

- *Fixed boundary* corresponds to the string end rigidly attached to the support (Fig. 5.7, left). Because the string end cannot move, the boundary condition is

$$x(\text{end}) = 0.$$

  In this case, the pulse is inverted, which means the pulse amplitude is preserved, but displacement is reversed. The inversion is understandable in terms of Newton's third law: Pulse produces upward force to the support, and the support exerts the opposite downward force to the string, leading to inversion.

- *Free boundary* corresponds to the string end loosely attached to the support, for example, tied to a light ring mounted on a smooth vertical post (Fig. 5.7, right). The string end can move freely in the perpendicular direction, and its end is always horizontal. Hence, the boundary condition is

$$\left.\frac{\partial x}{\partial z}\right|_{\text{end}} = 0.$$

  In this case, the pulse is not inverted, because the support cannot exert any force in the vertical direction.

Due to energy conservation, in the case of the perfect boundary, the amplitude of the reflected pulse is equal to the amplitude of the incident pulse. The pulse is totally reflected.

5.4.2 Transmission

We now consider the interaction of a pulse with an *imperfect boundary*. The imperfect boundary corresponds to the junction of two strings of different linear densities (Fig. 5.8). Note that perfect boundaries from Section 5.4.1 are just extreme cases of imperfect boundaries, for infinite and zero linear density.

Initially, there is a pulse in the primary string travelling towards the boundary. After reaching the junction, we get reflection as part of the pulse is bounced back and travels in the opposition direction, as explained in Section 5.4.1. However, part of the pulse passes the junction and travels in the same direction in the auxiliary string in a process called *transmission*.

The type of reflection and transmission depends on the properties of the auxiliary string:

1. If the auxiliary string has a *larger linear density* (Fig. 5.8, left), the reflected pulse is inverted and the transmitted pulse has a smaller amplitude and speed.

2. If the auxiliary string has a *smaller linear density* (Fig. 5.8, right), the reflected pulse is not inverted and the transmitted pulse has a larger amplitude and speed.

Due to the energy conservation, the sum of the energies of reflected and transmitted pulses must be equal to the energy of the incident pulse. By using boundary conditions we can calculate both amplitudes. However,

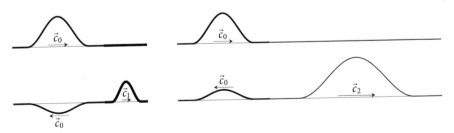

Figure 5.8: Reflection and transmission for two strings of different linear densities joined at their ends. On the left side of the junction is a primary string with the initial pulse, and on the right side of the junction is an auxiliary string. Two possibilities are presented with an auxiliary string of the larger linear density (left) and the smaller linear density (right). In both cases, part of the pulse is reflected and part is transmitted at the junction of both strings. Note that the pulse velocity in the strings differs (5.15). The calculation of amplitudes of reflected and transmitted pulses is beyond the scope of this book.

boundary conditions here are more complicated than the ones in the perfect boundary situation. Displacement and slope on both sides of the junction are nontrivially related. The process of calculation is beyond the scope of this book.

5.4.3 Superposition

Finally, we consider interaction between several pulses. Pulses and waves are solutions to the *linear differential equation*. Mathematically speaking, to this type of differential equation applies the principle of *superposition*; that is, if x_1 in x_2 are two arbitrary solutions, so is their linear combination $ax_1 + bx_2$, where a and b are arbitrary constants. In practice, if two pulses with wave functions $x_1(z, t)$ and $x_2(z, t)$ travel along some medium, the total wave function will be simply an algebraic sum of both wave functions $x_1(z, t) + x_2(z, t)$.

One of the consequences of superposition is that two pulses can travel one through another without destroying or changing themselves. Fig. 5.9 presents two examples of pulse superposition. Left pulse $x_1(z, t)$ travels to the right, and right pulse $x_2(z, t)$ travels to the left. When the waves overlap, the wave function is equal to the sum of both $x_1(z, t) + x_2(z, t)$. After splitting up, the pulses continue propagating in the initial direction. Wave forms remain unchanged, as if they never met.

A combination of two pulses or waves in the same area creates a resultant called *interference*. On the left side of Fig. 5.9, the displacements of both pulses have the same direction, so the amplitude of the resultant pulse is larger than amplitudes of the either individual pulse. This type of superposition is called *constructive interference*. On the right side of Fig. 5.9, the displacements of pulses have opposite directions, so the amplitude of the resultant pulse is smaller than the amplitude of the larger (and possibly the smaller) pulse. This type of superposition is called *destructive interference*.

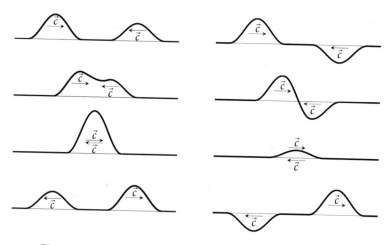

Figure 5.9: Constructive (left) and destructive (right) interference.

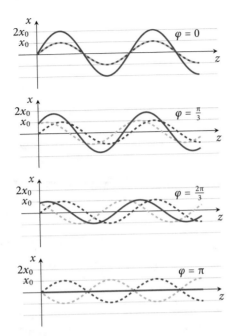

Figure 5.10: Interference of two identical harmonic travelling waves with different phases φ, propagating in the same direction. Individual waves are blue and green, and their interference is red. For $\varphi = 0$ (top), interference is fully constructive, and the amplitude of the interference is twice the amplitude of individual waves. For $\varphi = \pi$ (bottom), interference is fully destructive, and the two waves cancel out, resulting in zero interference.

5.5 Standing waves

In this section, we will limit ourselves to interactions between *identical harmonic travelling waves*, that is, waves of the same frequency/wavelength and same amplitude. We start with two identical waves that are propagating in the *same* direction with different phases φ. Assuming propagation in the right direction, their wave functions are

$$x_1(z, t) = x_0 \sin(kz - \omega t),$$
$$x_2(z, t) = x_0 \sin(kz - \omega t + \varphi).$$

The resultant wave or interference is then

$$x(z, t) = x_1(z, t) + x_2(z, t) = x_0\left[\sin(kz - \omega t) + \sin(kz - \omega t + \varphi)\right]$$
$$= 2x_0 \cos\left(\frac{\varphi}{2}\right) \sin\left(kz - \omega t + \frac{\varphi}{2}\right). \qquad (5.17)$$

The interference is presented in Fig. 5.10. We see that the interference of two identical travelling waves is itself a travelling wave because the wave function includes the $kz - \omega t$ part. The amplitude of the interference is $2x_0 \cos(\varphi/2)$, and the phase is $\varphi/2$. For phase $\varphi = 0$, amplitude of interference is $2x_0$, which is twice the amplitude of the individual waves. In

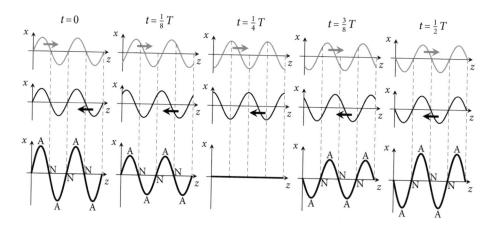

Figure 5.11: Interference of two identical waves propagating in opposite directions, one in the right direction (top) and one in the left direction (middle). The resultant wave function (bottom) corresponds to the standing wave. Nodes are labelled with 'N' and antinodes with 'A'.

that case, the crests and troughs of both waves coincide, so waves interfere constructively (Fig. 5.10, top). For phase $\varphi = \pi$, amplitude of interference is 0. The crests of one wave coincide with the troughs of the other, so the waves interfere destructively (Fig. 5.10, bottom), completely cancelling each other out. For situations with phases between those extreme values, the amplitude of interference ranges between 0 and $2x_0$ (Fig. 5.10, middle).

Next, we use the principle of superposition on two equal waves that are propagating in *opposite* directions. Their wave functions can be written as

$$x_1(z, t) = x_0 \sin(kz - \omega t),$$
$$x_2(z, t) = x_0 \sin(kz + \omega t).$$

The resultant wave or interference is then

$$x(z, t) = x_1(z, t) + x_2(z, t) = x_0\big[\sin(kz - \omega t) + \sin(kz + \omega t)\big]$$

$$x(z, t) = 2x_0 \cos(\omega t) \sin(kz). \tag{5.18}$$

The obtained equation represents the *standing harmonic wave* function.

The interference process is presented in Fig. 5.11. Unlike the previous situation (5.17), notice that the resultant wave is not travelling because wave function (5.18) lacks the $kz - \omega t$ part. Because spatial and temporal parts are separated, the interference consists of standing wave $2x_0 \sin(kz)$ that strengthen and weaken harmonically with function $\cos(\omega t)$. Therefore, all string fragments oscillate in phase (synchronous) but with different amplitudes.

From (5.18), we see that the oscillation amplitude of some points is equal to zero. This applies to points for which $\sin(kz) = 0$, that is,

$$kz = 0, \ \pi, \ 2\pi, \ 3\pi \ldots.$$

Combining with (5.9), we get

$$z = 0, \tfrac{1}{2}\lambda, \lambda, \tfrac{3}{2}\lambda \dots$$

Points with zero amplitude are called *nodes*.

On the other hand, from (5.18), we see that the maximum oscillation amplitude is $2x_0$. This applies to points for which $\sin(kz) = 1$, that is,

$$kz = \tfrac{1}{2}\pi, \tfrac{3}{2}\pi, \tfrac{5}{2}\pi, \tfrac{7}{2}\pi \dots$$

Combining with (5.9), we get

$$z = \tfrac{1}{4}\lambda, \tfrac{3}{4}\lambda, \tfrac{5}{4}\lambda, \tfrac{7}{4}\lambda \dots$$

Points with maximum amplitude are called *antinodes*.

Figure 5.11 uses 'N' for nodes and 'A' for antinodes. We observe that

- the distance between adjacent nodes is equal to $\lambda/2$,
- the distance between adjacent antinodes is equal to $\lambda/2$ and
- the distance between a node and an adjacent antinode is $\lambda/4$.

At first sight, interference of identical waves that are moving in the opposite direction seems to be an unlikely situation. However, standing waves are very common phenomena. They usually appear in the vicinity of (almost) perfect boundaries. From Section 5.4, we know that in this case, the reflected wave amplitude is equal to the incident wave amplitude. Therefore, *the standing wave usually appears as an interference of the incident wave and its reflection from the (almost) perfect boundary.* For a free boundary, the standing wave ends with an antinode, whereas for a fixed boundary, the standing wave ends with a node.

> **Info box**
>
> A standing wave usually appears as an interference of the incident wave and its reflection from the (almost) perfect boundary.

The most important example of standing waves concerns standing waves *between two boundaries*. Depending on the types of both boundaries, standing waves end with antinodes and/or nodes. With two boundary conditions, it is impossible to fit in an arbitrary standing wave. The possible standing waves are called *normal modes*. The first three normal modes for various boundary conditions are presented in Fig. 5.12. When there are two equal boundaries (Fig. 5.12, left and middle), wavelengths and frequencies of normal modes (5.10) are

$$\lambda_n = \frac{2L}{n},$$
$$f_n = n\frac{c}{2L}. \tag{5.19}$$

When there are two different boundaries (Fig. 5.12, right) wavelengths and frequencies of normal modes (5.10) are

$$\lambda_n = \frac{4L}{2n-1},$$
$$f_n = (2n-1)\frac{c}{4L}. \tag{5.20}$$

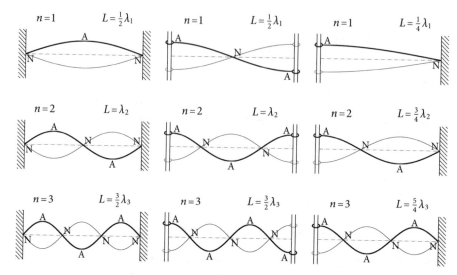

Figure 5.12: First three normal modes for a string ended with two perfect boundaries. Two fixed boundaries (left) lead to standing waves that end with
nodes on both sides, and free boundaries (middle) lead to standing
waves that end with antinodes on both sides. A combination of different boundaries (right) leads to standing waves that end with one node
and one antinode. Normal modes appear only for very particular relations between inter-boundary distance L and wavelength λ.

The frequency of the first normal mode, f_1, is called fundamental frequency. The frequencies of the other normal modes, f_n, $n \geq 2$, are integer
multiples of the fundamental frequency. Frequencies that exhibit such an
integer-multiple relationship form a *harmonic series*, so normal modes are
usually called harmonics. The fundamental frequency f_1 is the frequency
of the first harmonic, the frequency f_2 is the frequency of the second harmonic, and the frequency f_n is the frequency of the n-th harmonic.

This is a prominent effect in music. Regardless of the underlying physical mechanism, all instruments, including human voice, produce sound
in mediums with perfect boundaries, and fundamental frequency is always accompanied by higher harmonics. We therefore perceive all these
frequencies as the same tone. In particular, the interval between first two
harmonics in (5.19), that is, between particular frequency and its double, is
the basic musical interval *octave*. On the other hand, the ratio of amplitude
of the first harmonic to amplitudes of other harmonics are instrument dependent. This gives each instrument its distinctive tone colour or timbre.
We will elaborate on that in Section 6.2.5 on page 200.

Problems

5.1   Calculate the frequency of electromagnetic waves for wavelengths
of 0.50 µm and 10 µm. (6.0×10^{14} Hz, 3.0×10^{13} Hz)

5.2 Calculate the shortest distance between two parallel walls for which standing sound waves of frequency 110 Hz appear. Take the sound velocity to be 340 m/s. (1.54 m)

5.3 The ear canal is a tube of length 25 mm, open on one side and closed on the other side with the tympanic membrane. Calculate frequencies of all harmonics of the standing wave in the ear canal that can be heard. (3.4 kHz, 10.2 kHz, 17.0 kHz)

6 Sound propagation

Because acoustics (the study of sound) was traditionally developed independently of the other building physics subjects, several symbols have different meanings, even though the unit is the same. The most important differences are listed here to avoid confusion:

Symbol	Unit	Heat, Moisture, Light	Sound
A	m^2	area	equivalent absorption area
S	m^2	–	area
α	1	absorptance	absorbance
p	Pa	pressure of water vapour	sound pressure

Furthermore, whereas the study of heat and light use wavelength λ, the study of sound uses frequency f as a basic quantity to describe waves. Because both are simply related (5.10), they can be easily converted.

6.1 Introduction

Sound is a longitudinal mechanical wave. The adjective *mechanical* expresses the fact that the wave transmission is facilitated by oscillation of material particles (atoms, molecules). Without the presence of material particles, in a vacuum, sound cannot propagate.

Oscillation of an individual particle occurs on a microscopic scale, so tracking its movement is an infeasible task. On the other hand, collective oscillations of particles create regions with larger density and regions with smaller density (Fig. 6.1). For fluids, this can be conveniently described

 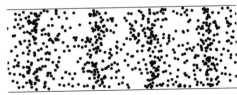

Figure 6.1: A sketch of the gas particles layout in case of silence (left) and sound (right).

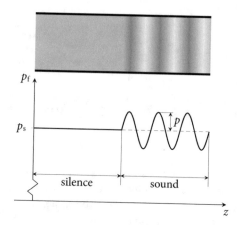

Figure 6.2: Spatial variation of fluid pressure for silence and sound.

by pressure (1.12), that is, regions with larger and smaller pressure. Fluid pressure is a macroscopic phenomenon that we will use to describe sound waves.

Spatial variation of fluid pressure is shown in Fig. 6.2. Without the presence of the sound waves, fluid pressure p_f is constant everywhere and amounts to *static pressure* p_s ($p_s = 1.013 \times 10^5$ Pa for air at sea level). The presence of sound waves changes fluid pressure, but only for a small fraction of the static pressure (smaller by at least 10 orders of magnitude). That small pressure change is called *sound pressure* p (Pa). Total fluid pressure is therefore

$$p_f = p_s + p. \tag{6.1}$$

The sound pressure is especially convenient, because it can be measured easily. The basic instrument for sound measurement—a *microphone*—is depicted in Fig. 6.3. The essential part of any sound-detecting instrument is a membrane, which is the *diaphragm* in a microphone and the *ear drum* in an ear. Pressure on the outer side of the membrane is equal to $p_s + p$, and pressure on the inner side of the membrane is p_s. The net force to the

diaphragm

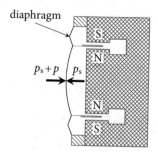

Figure 6.3: Sketch of a microphone. The most essential part is the diaphragm, a thin, semi-rigid membrane. The diaphragm moves due to sound pressure and is attached to the coil that produces an electrical signal by moving in a magnetic gap. The principle of a loudspeaker is just the opposite in that an electrical signal moves the coil and the diaphragm produces sound pressure.

membrane of area S as

$$F = (p_s + p) S - p_s S = p S$$

is directly proportional to the sound pressure. Net force produces membrane movement whose displacement is proportional to sound pressure. Membrane displacement is then transformed into an electrical or nerve signal.

6.1.1 Wave function

To understand fluid (liquid and gas) dynamics in the presence of sound, let's observe a small fluid fragment within a tube of cross-sectional area S. In the static situation (Fig. 6.4, top), the length of the fluid fragment is Δz, and the volume is $V = S\Delta z$. Because no sound is present, pressures both inside and outside of the fluid fragment is equal to static pressure p_s.

In the dynamic situation (Fig. 6.4, bottom), the fluid fragment pressure increases by sound pressure to $p_s + p$. The fragment also shifts from the equilibrium position so that the left-side displacement is x_1 and the right-side displacement is x_2. The new fluid fragment volume is equal to $S(\Delta z + x_2 - x_1) = S(\Delta z + \Delta x)$, so the volume change is $\Delta V = S\Delta x$. If we use the definition of the modulus of compression K as

$$\Delta p_f = -K\frac{\Delta V}{V},$$

and observe that the change in fluid pressure is simply sound pressure, we get

$$p = -K\frac{\Delta V}{V} = -K\frac{S\Delta x}{S\Delta z}.$$

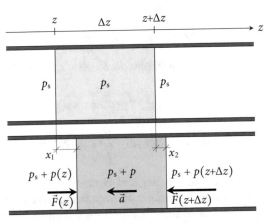

Figure 6.4: Small fluid fragment within a tube of cross-sectional area S in the static (top) and the dynamic situation (bottom). The fluid fragment volume changes because displacement on the left side (x_1) is different from displacement on the right side (x_2).

For the infinitesimally thin fluid fragment, the relation between pressure and displacement is

$$p = -K\frac{\partial x}{\partial z}.$$ (6.2)

This is an extremely important relation because it connects the macroscopic quantity (sound pressure) with the microscopic quantity (particle displacement).

Because pressure on both sides of the fragment are no longer the same in the dynamic situation (Fig. 6.4, bottom), the difference of forces causes fluid fragment acceleration a. We can calculate acceleration using Newton's second law as

$$F(z) - F(z+\Delta z) = S[p(z) - p(z+\Delta z)] = ma.$$

Because the pressure is a smooth function of position z, we can relate two pressures using the Taylor series as

$$p(z+\Delta z) = p(z) + \frac{\Delta z}{1!}\frac{\partial p}{\partial z} + \frac{\Delta z^2}{2!}\frac{\partial^2 p}{\partial z^2} + \cdots \approx p(z) + \Delta z\frac{\partial p}{\partial z}$$

to get

$$-S\Delta z\frac{\partial p}{\partial z} = -V\frac{\partial p}{\partial z} = -\frac{m}{\rho}\frac{\partial p}{\partial z} = ma.$$

Inserting the relation between pressure and displacement (6.2) and definition of acceleration (5.12), we get

$$\frac{\partial^2 x}{\partial t^2} = \frac{K}{\rho}\frac{\partial^2 x}{\partial z^2}.$$

Comparing the expression with the wave equation (5.16), we conclude that the solution is a sound wave (5.7) as

$$p(z, t) = p_0 \sin(\omega t - kz),$$

where p_0 is the sound pressure amplitude. *Speed of sound* in fluids (liquids and gases) is therefore

$$c = \sqrt{\frac{K}{\rho}}.$$ (6.3)

In the most important liquid, water ($K = 2.1 \times 10^9 \text{ N/m}^2$), speed of sound is 1.4 km/s.

For gases, fragments are contracting and expanding adiabatically, that is, without heat exchange (Section 1.9.2 on page 23). Therefore, the expression can be simplified using the expression for the modulus of compression for the adiabatic processes (1.41), ideal gas law (1.12) and amount of substance (1.9) to get

$$c = \sqrt{\frac{\gamma RT}{M}},$$ (6.4)

where γ is ratio of heat capacities. If we use typical values for air of $\gamma = 1.4$ and $M = 0.029\,\text{kg/mol}$, the expression can be transformed into the form defined by standard ISO 9613-1 [31] as

$$c = 343.2\,\frac{\text{m}}{\text{s}} \cdot \sqrt{\frac{T}{T_0}}, \tag{6.5}$$

where $T_0 = 293\,\text{K}$. The speed of sound at $20\,°\text{C}$ is therefore $343.2\,\text{m/s}$.

In a similar procedure, we can show that the speed of sound in solids is equal to

$$c = \sqrt{\frac{E}{\rho}}, \tag{6.6}$$

where E (N/m^2) is the modulus of elasticity. For example, the speed of sound in iron ($E = 2.1 \times 10^{11}\,\text{N/m}^2$, $\rho = 7900\,\text{kg/m}^3$) is $5.2\,\text{km/s}$.

We can define another relation that connects macroscopic quantity (sound pressure) with microscopic quantity (particle velocity). However, this relation depends on the wave type.

For a purely harmonic *travelling sound wave*, movement of individual particles is described by

$$x = x_0 \sin(\omega t - kz),$$

$$v = \frac{dx}{dt} = x_0 \omega \cos(\omega t - kz),$$

and the expression for sound pressure can be obtained using (6.2):

$$p = -K\frac{dx}{dz} = Kx_0 k \cos(\omega t - kz).$$

We see that particle displacements and sound pressure are out of phase in time and space. On the other hand, particle velocities and sound pressure are in phase (synchronised), and their quotient can be calculated using (5.6) and (6.3):

$$\frac{p}{v} = \frac{Kk}{\omega} = \frac{K}{c^2}c$$

$$\frac{p}{v} = \rho c. \tag{6.7}$$

For a purely harmonic *standing sound wave*, the movement of particles is described by (5.18) as

$$x = x_0 \cos(kz) \sin(\omega t),$$

$$v = \frac{dx}{dt} = x_0 \omega \cos(kz) \cos(\omega t),$$

and the expression for sound pressure can be obtained using (6.2) as

$$p = -K\frac{dx}{dz} = Kx_0 k \sin(kz) \sin(\omega t).$$

We see that particle displacements and sound pressure are out of phase in space and in phase for time. On the other hand, particle velocities and sound pressure are out of phase in time and space, so expression (6.7) can be used only for the quotient of their *amplitudes*:

$$\frac{p_0}{v_0} = \rho c. \tag{6.8}$$

6.1.2 Power and energy transfer

Sound intensity

The essential property of sound waves is their ability to transfer energy. As stated in the Introduction, we will define the density of energy flux rate called *sound intensity* i (W/m^2). Sound intensity is the quotient of mechanical energy dE transferred through a surface by the product of surface area S and transfer time dt:

$$i = \frac{dE}{S\,dt} = \frac{P}{S}. \tag{6.9}$$

The amount of transferred energy is equal to the amount of work done on the imaginary surface by fluid particles of velocity v:

$$P = F v = p_f S v. \tag{6.10}$$

Combining the preceding equations and (6.1) we get

$$i = p_f v = (p_s + p)v. \tag{6.11}$$

Sound pressure, particle velocities and consequently sound intensity are changing rapidly—oscillating with rather large sound frequency—so temporary sound intensity is neither easily observable nor a very useful quantity. We are therefore more interested in *time-averaged sound intensity* I (W/m^2):

$$I = \langle i \rangle = \frac{1}{t'} \int_0^{t'} i\,dt. \tag{6.12}$$

Taking into account that the average particle velocity is $\langle v \rangle = 0$, we can simplify the expression for sound intensity to

$$I = \langle p_s v \rangle + \langle pv \rangle = p_s \langle v \rangle + \langle pv \rangle = \langle pv \rangle. \tag{6.13}$$

For a *travelling sound wave*, by virtue of (6.7) the value of time-averaged sound intensity is

$$I = \frac{\langle p^2 \rangle}{\rho c}. \tag{6.14}$$

The square root of time average

$$\langle p^2 \rangle = \frac{1}{t'} \int_0^{t'} p^2 \, dt = p_{rms}^2$$

is usually called *root-mean-square sound pressure* p_{rms}, so time-averaged sound intensity is usually recorded as

$$I = \frac{p_{rms}^2}{\rho c}. \tag{6.15}$$

For a *standing sound wave*, the value of time-averaged sound intensity is

$$I = 0, \tag{6.16}$$

as $\langle \sin(\omega t) \cos(\omega t) \rangle = 0$. This is understandable because in standing waves, energy only oscillates, whereas the net energy transfer through the space is equal to zero.

It is common to omit the 'time average' attribute for sound intensity, and we will do the same in the rest of the book.

Note that we were not time-averaging the density of heat flow rate. Time-averaging the density of energy flux rate is crucial only for energy transferred by waves.

The importance of sound intensity is not limited to the description of energy transfer. Its value directly corresponds to the perceived or measured sound loudness. As we noted previously, the essential part of a sound-detecting instrument is the membrane (Fig. 6.3 on page 186). The power intercepted by the membrane is obtained by multiplying the sound intensity and membrane area S:

> **Info box**
> Sound intensity is a basic measure of sound loudness.

$$P = I \, S. \tag{6.17}$$

Thus, sound intensity is a measure of *sound loudness*.

Another useful quantity is time-averaged density of sound energy e (J/m^3):

$$e = \left\langle \frac{dE}{dV} \right\rangle. \tag{6.18}$$

We will show that for travelling waves, this quantity is directly connected to sound intensity. Let's observe a small wave fragment of length dz within a tube of cross-sectional area S just before and just after it passes certain location z (Fig. 6.5). The volume of the fragment is

$$dV = S \, dz. \tag{6.19}$$

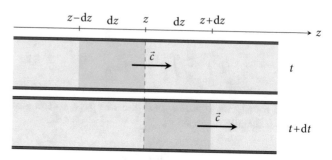

Figure 6.5: Small wave fragment of length dz within a tube of cross-sectional area S moving to the right just before (top) and just after (bottom) passing location z. During the time period dt, the sound energy transferred at location z is equal to the sound energy of the wave fragment.

In time dt, the wave fragment passes the location z, so the transferred sound energy at this location is equal to the sound energy of fragment dE. Using (6.9) and (6.12), we get

$$I = \left(\frac{dE}{S\,dt}\right) = \left(\frac{dE}{dV}\frac{dz}{dt}\right). \tag{6.20}$$

In time dt, the wave fragment is shifted for dz, so the time derivative of location dz/dt is the time-independent speed of sound c:

$$I = c\,e. \tag{6.21}$$

Sound intensity is the product of the speed of sound and the density of sound energy. The expression is very useful, because it is valid for all energy transfers due to wave propagation and not only for sound waves. In particular, the time-averaged intensity is always a product of the time-averaged density of wave energy and the speed of waves.

Sound power

Info box

Sound power is a basic measure of sound emission.

To characterise sound sources, we also introduce the *sound power* quantity with unit *watt* P (W) as

$$P = \frac{dE}{dt}, \tag{6.22}$$

where E is the emitted sound energy, and t is the time. Note that it is implicitly assumed that sound power is time-averaged, in line with time-averaged intensity.

The sound power of common sound emitters such as speakers is only a small fraction of their rated (electrical) power. This means that only a small fraction of provided electrical energy is transferred to sound energy, whereas the rest is dissipated as heat.

6.1.3 Logarithmic description of sound power and intensity

The sound power of common sound sources spans over several orders of magnitude, from 1×10^{-11} W for breathing to 10 W for a symphonic orchestra. Similarly, normal human hearing is also capable of processing sound intensity of several orders of magnitude without permanent harm, from about 1×10^{-12} W/m^2 for threshold of hearing to about 1 W/m^2 for a loud concert. We will elaborate more on human ear limitations in Section 6.3.2.

To describe such a wide range of sound powers and intensities, we represent the quantities on a logarithmic scale. In acoustics, as well as in other engineering disciplines, it is common to define *level* with the unit *bel* L (B) as

$$L_Q = \lg \frac{Q}{Q_0},$$

where $\lg = \log_{10}$ is the decimal logarithm, Q is the arbitrary quantity and Q_0 is its reference value. The basic unit bel (B) is too coarse, so levels are usually defined and expressed in the 10 times smaller unit *decibel* (dB) by rewriting the common definition as

$$L_Q = 10 \lg \frac{Q}{Q_0}.$$

Hereafter, we will always assume that dB is the basic unit of any level.

First, we use this convention to define *sound power level* L_W (dB) as

$$L_W = 10 \lg \frac{P}{P_0}, \tag{6.23}$$

where $P_0 = 1.0 \times 10^{-12}$ W is the sound power reference value. Similarly, we define *sound intensity level* as

$$L_p = 10 \lg \frac{I}{I_0}, \tag{6.24}$$

where $I_0 = 1.0 \times 10^{-12}$ W/m^2 is the sound intensity reference value. Sound intensity and sound intensity level are usually determined by pressure measurements, so from (6.15), we get

$$L_p = 10 \lg \frac{p_{rms}^2}{p_0^2} = 10 \lg \left(\frac{p_{rms}}{p_0} \right)^2 = 20 \lg \frac{p_{rms}}{p_0},$$

$$L_p = 20 \lg \frac{p_{rms}}{p_0}, \tag{6.25}$$

Table 6.1: Typical activities associated with some sound pressure levels.

Activity/Place	Sound Pressure Level
threshold of hearing	0 dB
clock ticking	20 dB
quiet room	40 dB
normal conversation	60 dB
city traffic	80 dB
chainsaw	100 dB
rock concert	120 dB
gunshot	140 dB

where $p_0 = \sqrt{\rho c I_0} = 2.0 \times 10^{-5}$ Pa is the sound pressure reference value. Despite essentially describing sound intensity, L_p (dB) is called the *sound pressure level* and is defined in terms of the sound pressure reference value.

The obvious benefit of levels is simple articulation. The sound power level for breathing is 10 dB and for a symphonic orchestra is 130 dB. Sound intensity for the threshold of hearing is 0 dB and for a loud concert is 120 dB. Typical activities associated with some sound pressure levels are listed in Table 6.1.

Another viable advantage is a more intuitive description of the sensation. It is well known that the magnitude of human sensation is not directly proportional to the intensity of the stimulus. The exact relationship is still debatable, but one prediction, the Weber-Fechner law, states that the magnitude of human sensation is proportional to the logarithm of the intensity of the stimulus.

The disadvantage of levels is that they are essentially dimensionless, although their unit is decibel. This means that all four level quantities in this book, the sound power level, the linear sound power level, the surface sound power level and the sound pressure level, all have the same unit. As a result, we can no longer identify the quantity from the unit and must be much more careful in our calculations.

Another disadvantage is that levels are auxiliary quantities without direct physical meaning, so they cannot be directly measured and mathematical operations on them are unintuitive and complicated.

Let's start with doubling the quantity value:

$$L(2Q) = 10 \lg \frac{2Q}{Q_0} = 10 \lg 2 + 10 \lg \frac{Q}{Q_0} = L(Q) + 3 \,\text{dB}. \qquad (6.26)$$

We see that duplication of sound power or intensity leads to an increase of the corresponding level by 3 dB. In fact, each additional duplication leads to an additional increase by 3 dB in the corresponding levels: Quadruplication means an increase by 6 dB and octuplication means an increase by 9 dB. On the other hand, each halving of sound power or intensity means a decrease by 3 dB in the corresponding levels.

We now proceed with establishing important level calculation procedures. The basic operation is addition of quantity values. We start by adding two values:

$$Q = Q_1 + Q_2.$$

In this case, we get

$$L = 10\lg\frac{Q}{Q_0}, \quad L_1 = 10\lg\frac{Q_1}{Q_0}, \quad L_2 = 10\lg\frac{Q_2}{Q_0},$$

$$Q = Q_0\, 10^{0.1L}, \quad Q_1 = Q_0\, 10^{0.1L_1}, \quad Q_2 = Q_0\, 10^{0.1L_2}$$

$$\implies L = 10\lg\left(10^{0.1L_1} + 10^{0.1L_2}\right).$$

The equation can be generalised for n quantity values:

$$Q = \sum_{i=1}^{n} Q_i,$$

$$L = 10\lg\left(\sum_{i=1}^{n} 10^{0.1L_i}\right). \tag{6.27}$$

Next, we average n quantity values:

$$Q = \frac{1}{n}\sum_{i=1}^{n} Q_i,$$

$$L = 10\lg\left(\frac{1}{n}\sum_{i=1}^{n} 10^{0.1L_i}\right). \tag{6.28}$$

The last important operation is multiplication of the quantity value with the dimensionless number as

$$Q_N = N Q_1,$$
$$L_N = L_1 + 10\lg N, \tag{6.29}$$

of which (6.26) is a special case.

6.2 Sound sources and their properties

In Section 6.1.3 we defined auxiliary (level) quantities for sound power and intensity (dashed lines):

	source	receiver
quantity	$P\,(\mathrm{W}) \longrightarrow$	$I\,(\mathrm{W/m^2})$
level	$L_W\,(\mathrm{dB}) \longrightarrow$	$L_p\,(\mathrm{dB})$

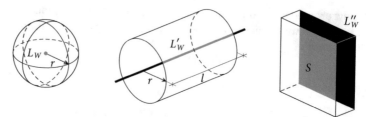

Figure 6.6: Sound propagation in the case of an isotropic point sound source (left), isotropic linear sound source (middle) and isotropic surface sound source (right). Sound sources and corresponding surfaces with constant sound intensity are shown.

In this section, we will make a connection between sound sources and sound receivers, which are represented by sound power and intensity, respectively (full lines). In order to find the connection, we will elaborate on how sound waves propagate from the sound source to the sound receiver. We will study the simplest situations under the following idealised conditions:

- *A free sound field* is a region without sound reflections, that is, without adjacent reflecting surfaces. In most practical situations, this is not feasible; However, it is assumed that the free sound field conditions are met if the sound pressure level directly from the sound source is 6 dB or preferably 10 dB greater than the level caused by reflections.

- *Isotropic sound sources* emit sound (as well as energy) evenly in all directions. In Section 6.2.4, we will also study a few simple situations with anisotropic sound sources.

6.2.1 Isotropic point sound source

Let's assume that we have an isotropic point sound source with constant sound power P. We must figure out for which part of the space the sound intensity is constant. For symmetry reasons, sound intensity depends only on the distance from the source, so the sound intensity should be constant for a spherical surface centred in the sound source (Fig. 6.6, left). Note that all sound energy that leaves the source must penetrate that spherical surface. Therefore, we can calculate sound intensity by dividing sound power with the area of the surface (6.9) as

$$I = \frac{P}{S} = \frac{P}{4\pi r^2},$$

(6.30)

where S is the area of spherical surface, and r is the distance from the sound source.

We could also come to the same conclusion with a more tangible procedure. As shown in Fig. 5.1 on page 169, agitation at one point on the water surface forms circular wave fronts. As a wavefront moves away from this point, its length increases, but its crest becomes lower. This is due to the fact that the energy and power—energy divided by the period of oscillation—of a single wavefront are constant, which means that energy and power per length of a wavefront should decrease.

Similarly, a point sound source in space forms spherical wavefronts. As a single wavefront moves away from the source, its area increases, which means that the energy and power per area of a wavefront decrease. The intensity is obtained by dividing the source sound power by the wavefront area corresponding to the area of the sphere.

After we have determined the relationship between sound power and sound intensity, we continue with the determination of the relationship between sound power level and sound pressure level. Putting the former relationship in (6.24), we obtain for the sound pressure level

$$L_p = 10 \lg \frac{I}{I_0} = 10 \lg \left(\frac{P}{4 \pi r^2} \frac{r_0^2}{P_0} \right) = 10 \lg \left(\frac{P}{P_0} \frac{r_0^2}{4 \pi r^2} \right)$$

$$= 10 \lg \frac{P}{P_0} - 10 \lg \frac{r^2}{r_0^2} - 10 \lg 4\pi,$$

$$L_p = L_W - 20 \lg \frac{r}{r_0} - 11 \, \text{dB}, \tag{6.31}$$

where $r_0 = 1 \, \text{m}$ is the distance reference value. By doubling the distance from the source $r \rightarrow 2r$, the sound pressure level reduces by $20 \lg 2 = 6 \, \text{dB}$.

Note that the expression for the point sound source (6.31) is also valid for finite sound sources as long as the distance between the sound source and the observation position is much larger than the dimensions of the sound source.

6.2.2 Isotropic linear sound source

Let's assume that we have an isotropic *infinite* linear sound source of constant linear sound power P' (W/m) as

$$P' = \frac{P}{l}, \tag{6.32}$$

where P is the emitted sound power by the sound source section of length l, and of constant *linear sound power level* L'_W (dB) as

$$L'_W = 10 \lg \frac{P'}{P'_0}, \tag{6.33}$$

where $P'_0 = 1 \times 10^{-12}$ W/m is the linear sound power reference value.

We must figure out for which part of the space the sound intensity is constant. For symmetry reasons, sound intensity depends only on the distance from the source, so the sound intensity should be constant for an infinite cylindrical surface centred around the sound source (Fig. 6.6, centre).

Note that all sound energy that leaves the source must penetrate that cylindrical surface. Therefore, we can calculate sound intensity by dividing the sound power of the sound source section of length l with the corresponding area of the surface (6.9) as

$$I = \frac{P}{S} = \frac{P'l}{2\pi rl} = \frac{P'}{2\pi r}, \tag{6.34}$$

where S is the area of cylindrical surface, and r is the distance from the sound source.

Putting that into (6.24), for sound pressure level we get

$$L_p = 10 \lg \frac{I}{I_0} = 10 \lg \left(\frac{P'}{2\pi r} \frac{r_0}{P'_0} \right) = 10 \lg \left(\frac{P'}{P'_0} \frac{r_0}{2\pi r} \right)$$

$$= 10 \lg \frac{P'}{P'_0} - 10 \lg \frac{r}{r_0} - 10 \lg 2\pi,$$

$$L_p = L'_W - 10 \lg \frac{r}{r_0} - 8 \, \text{dB}. \tag{6.35}$$

By doubling distance from the source $r \to 2r$, the sound pressure level reduces by $10 \lg 2 = 3 \, \text{dB}$. This reduction is half as large as for the point sound source, which means that the linear sound sources' impact diminishes slower.

The expression for the linear sound source (6.35) is also valid for a finite linear sound source as long as the distance between the observation point and the centre of the sound source is *much smaller* than the length of the sound source. On the other hand, if this distance is *much larger* than the length of the sound source, such a source can be treated as a point sound source (6.31) with the sound power level obtained using (6.29) as in

$$L_W = L'_W + 10 \lg \frac{l}{l_0}, \tag{6.36}$$

where $l_0 = 1 \, \text{m}$.

6.2.3 Isotropic surface sound source

Let's assume that we have an isotropic *infinite* flat surface sound source of constant surface sound power P'' (W/m) as

$$P'' = \frac{P}{S}, \tag{6.37}$$

where P is the emitted sound power by the sound source section of area S, and of constant *surface sound power level* L''_W (dB) as

$$L''_W = 10 \lg \frac{P''}{P''_0}, \tag{6.38}$$

where $P''_0 = 1 \times 10^{-12} \, \text{W/m}^2$ is the surface sound power reference value.

Note that the sound intensity and sound pressure level are distance independent (Fig. 6.6, right):

$$I = \frac{P}{S} = \frac{P''S}{S} = P''.$$
(6.39)

Putting that into (6.25), for sound pressure level we get

$$L_p = 10\lg\frac{I}{I_0} = 10\lg\frac{P''}{P''_0},$$

$$L_p = L''_W.$$
(6.40)

The expression for a surface sound source (6.40) is valid for finite surface sound sources as long as the distance between the observation point and the centre of the sound source is *much smaller* than the dimensions of the sound source. On the other hand, if this distance is *much larger* than the dimensions of the sound source, such a source can be treated as a point sound source (6.31) with the sound power level obtained using (6.29) as in

$$L_W = L''_W + 10\lg\frac{S}{S_0},$$
(6.41)

where $S_0 = 1\,\mathrm{m}^2$.

6.2.4 Directivity correction

When we calculated sound intensity (6.30) and sound pressure level (6.31) from the point sound source, we assumed that the sound source was isotropic. Most real sound sources, however, emit sound only in certain directions. We must therefore divide the sound power only by the part of the sphere surface into which sound propagates, transforming (6.30) into

$$I = \frac{P}{S} = \frac{P}{\Omega r^2},$$

where Ω is the *solid angle* (8.16), defined in Section 8.3.2 on page 249. For example, $\Omega = 4\pi$ for the whole space; $\Omega = 2\pi$ for the half-space; $\Omega = \pi$ for the quarter space; and $\Omega = \frac{1}{2}\pi$ for the eighth space (Fig. 6.7). Instead of using the preceding expression, it is more common to account for space limited sound emission by adding *directivity correction* D_C (dB) to the sound pressure level as in

$$L_p = 10\lg\frac{P}{\Omega r^2} = 10\lg\left(\frac{P}{4\pi r^2}\frac{4\pi}{\Omega}\right) = L_{p0} + D_C,$$

$$D_C = 10\lg\frac{4\pi}{\Omega},$$

where L_{p0} is the sound pressure level of the corresponding isotropic point sound source.

For the half-space, sound intensity doubles, and $D_C = 3\,\mathrm{dB}$; for the quarter space, sound intensity quadruples, and $D_C = 6\,\mathrm{dB}$; and for the eighth space, sound intensity octuples, and $D_C = 9\,\mathrm{dB}$ (Fig. 6.7).

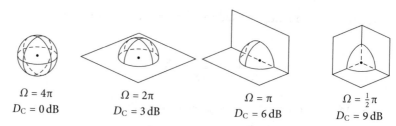

$$\Omega = 4\pi \qquad\qquad \Omega = 2\pi \qquad\qquad \Omega = \pi \qquad\qquad \Omega = \tfrac{1}{2}\pi$$
$$D_C = 0\,\text{dB} \qquad D_C = 3\,\text{dB} \qquad D_C = 6\,\text{dB} \qquad D_C = 9\,\text{dB}$$

Figure 6.7: Directivity correction. If sound propagation is limited to a smaller solid angle Ω for the same sound power, the sound pressure level increases for the directivity correction D_C.

Directivity correction is also used for isotropic or quasi-isotropic sound sources that are positioned near highly reflective surfaces, such as hard smooth surfaces. For example, if the source is positioned near one such surface, all sound is reflected from it, and no sound energy is lost. This means that complete sound power propagates into the half-space, and $D_C = 3\,\text{dB}$. Similarly, for a sound source near the edge of two highly reflective perpendicular surfaces, $D_C = 6\,\text{dB}$, and for a sound source near the corner of three highly reflective perpendicular surfaces, $D_C = 9\,\text{dB}$. Most hard, smooth surfaces are highly reflective. Sound reflections will be elaborated on in Section 7.1 on page 217.

6.2.5 Spectral characteristics

Different sound sources produce different types of the sounds. For example, a tuning fork produces the sound of a single frequency called *tone*. The function of tone is a perfect sine curve (Fig. 6.8, left). On the other hand, as discussed in Section 5.5 on page 179, instruments also produce higher harmonics, creating *timbre*. The function is still repetitive, but not a perfect sine curve (Fig. 6.8, middle). Nonmusical sound sources usually produce sounds of a continuous and broad spectrum, creating random

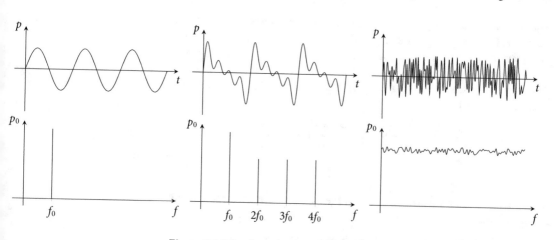

Figure 6.8: Waveform (top) and frequency spectrum (bottom) for a tone (left), timbre (middle) and noise (right).

function (Fig. 6.8, right). This type of sound is technically called *noise*; however, noise has multiple meanings. In Section 6.3.1 we will also become acquainted with the broader definition.

6.3 Human perception of sound and sound rating

6.3.1 Noise

So far, we have been only concerned with sound propagation. Yet, our primary goal is to evaluate the influence of sound on human beings and to establish conditions in which negative influences have to be prevented.

The term *noise* is, in its broader definition, used to describe any unwanted and possibly harmful sound. The designation of a particular sound as an *unwanted sound* is a matter of subjective opinion based on individual preferences. In fact, just any sound may be regarded unwanted by at least a small group of people. To maintain neutrality, most modern legislations consider essentially any sound produced by humans and their devices as a noise and tend to control its quantity.

Noise has several well documented and proved harmful effects. Among *physiological effects*, there are hearing impairment, hypertension, cardiovascular problems and sleep disturbance. In addition, *psychological effects* include creating stress, increasing workplace accident rates and stimulating the aggression and other antisocial behaviours. The negative effects of the noise pose huge social costs, so the efforts to study and reduce noise quantity are increasing.

However, despite all the studies, there is still no well-established consensus under which conditions and at which values noise becomes unacceptable. Determination of noise limit values is therefore still the sovereign right of state regulations. Nevertheless, we will discuss the *relative effect* of various circumstances to noise-produced annoyance and damage and present good practices in noise regulation.

6.3.2 Physiological perception

We start by considering the physiological properties of human hearing, with an emphasis on sound intensity and frequency.

We have already mentioned in Section 6.1.3 that normal human hearing is capable of processing sound intensity of several orders of magnitude. The smallest audible sound intensity, also called the *threshold of hearing*, amounts to about 1×10^{-12} W/m^2, whereas the largest safe sound intensity, also called the *threshold of pain*, amounts to about 1 W/m^2. From (6.24), we find that the corresponding sound pressure levels are 0 dB and 120 dB for threshold of hearing and threshold of pain, respectively.

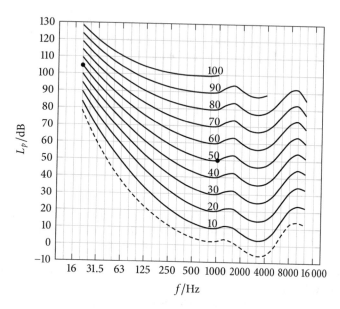

Figure 6.9: Isophones (equal-loudness contours) for different phon values (full lines) and absolute threshold of hearing (dashed line) [21]. A 105 dB sound at 20 Hz and a 50 dB sound at 1 kHz lay at the same isophone, as they are perceived as equally loud.

Similarly, from (6.15), we find that the corresponding rms sound pressures are 2.0×10^{-5} Pa and 2.0 Pa for threshold of hearing and threshold of pain, respectively.

On the other hand, normal human hearing is only capable of processing frequency range from 20 Hz to 20 kHz. Sound of frequencies smaller than 20 Hz is called infrasound and is used for earthquake monitoring, whereas sound of frequencies larger than 20 kHz is called ultrasound and is used for nondestructive examination of living beings and solid objects.

However, even within the audible range, equal sound intensities are not perceived as equally loud. For example, a 105 dB sound at 20 Hz is perceived by humans as equally loud as 50 dB sound at 1 kHz. To describe frequency sensitivity, we define *phon* as the sound pressure level at 1 kHz that has the same perceived loudness as a particular sound. For example, 105 dB at 20 Hz is equal to 50 phon, because it is perceived as loud as a 50 dB, 1 kHz sound. The effect is usually presented in terms of *isophones* (equal-loudness contours), which are the lines that connect points with the same phon value (Fig. 6.9).

Isophones are defined in standard ISO 226 [21]. The standard first divides the audible range into one-third octave bands, designated by mid-band frequencies 10 Hz, 12.5 Hz, <u>16 Hz</u>, 20 Hz, 25 Hz, <u>31.5 Hz</u>, 40 Hz, 50 Hz, <u>63 Hz</u>, 80 Hz, 100 Hz, <u>125 Hz</u>, 160 Hz, 200 Hz, <u>250 Hz</u>, 315 Hz, 400 Hz, <u>500 Hz</u>, 630 Hz, 800 Hz, <u>1 kHz</u>, 1.25 kHz, 1.6 kHz, <u>2 kHz</u>, 2.5 kHz, 3.15 kHz, <u>4 kHz</u>, 5 kHz, 6.3 kHz, <u>8 kHz</u>, 10 kHz, 12.5 kHz, <u>16 kHz</u>, 20 kHz. Octave mid-band frequencies are the ones underlined. The standard then defines isophone sound pressure level values at those frequencies.

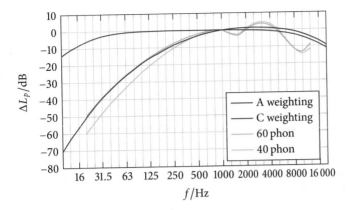

Figure 6.10: A and C weighting functions drawn together with reversed 40 and 60 isophones, centred at 1 kHz, 0 dB [20].

In order to compensate for the effect of the frequency-dependent sensitivity of human hearing, the result of measurement is adjusted using weighting functions, for example, A weighting function ΔL_{pA}:

$$L_{pA} = L_p + \Delta L_{pA}. \tag{6.42}$$

Standard IEC 61672-1 defines two weighting functions, displayed in Fig. 6.10, of which A is used by default, and C is used only for special applications, such as peak sound measurements. In fact, the A weighting function fits well with the loudness of 60 phon, but it is less appropriate for other loudnesses.

When particular frequency weighting is applied, this is indicated by the corresponding subscript, for example, L_{pA}. Frequency weighting is more commonly, albeit incorrectly, indicated by the addition of a postscript to a unit, for example, dB(A) or dBA. An example of A weighting is displayed in Fig. 6.11.

> **Info box**
>
> Weighting function (most prominently A weighting) compensate for frequency-dependent sensitivity of human hearing.

6.3.3 Psychological perception

In this section, we turn our attention to the psychological properties of human hearing. Instead of the objective term *sound*, we now turn to the term *noise*. It turns out that the noise of the same frequency and sound pressure level creates various degrees of annoyance due to human brain sound processing. This phenomenon is accounted for by introducing *level adjustments K* (dB).

The most important conditions that influence the psychological perception of noise are

- period of the day,
- location,
- presence of tones and
- presence of impulsive noise.

Period of the day

The physiologically same amount of noise is most annoying during the night, when people rest, and least annoying during the day, when people are active. In order to account for this sensitivity, standard ISO 1996-1 [25] defines *rating levels* for different day periods. Day is split into a daytime period (d hours), an evening period (e hours) and a night period (n hours). For each day period, rating levels are determined by averaging sound pressure levels measured or calculated within the particular period: daytime rating level L_{Rd}, evening rating level L_{Re} and night rating level L_{Rn}. The standard also introduces composite whole-day rating levels, called day-evening-night rating level as

$$L_{Rden} = 10 \lg \left[\frac{d}{24} 10^{0.1 L_{Rd}} + \frac{e}{24} 10^{0.1 (L_{Re} + K_e)} + \frac{24 - d - e}{24} 10^{0.1 (L_{Rn} + K_n)} \right],$$

where the level adjustment for evening period K_e and the level adjustment for night period K_n account for a larger annoyance in those two time periods. Note that without adjustment levels, the preceding expression would simply represent the average daily sound pressure level (6.28). The standard recommended values are K_e = 5 dB and K_n = 10 dB [25].

Directive 2002/49/EC of the European Parliament [61] calls the rating levels 'noise indicators'. The directive's recommended default time periods are 07:00 to 19:00 for the daytime period (d = 12 h), 19:00 to 23:00 for the evening period (e = 4 h) and 23:00 to 07:00 local time for the night period (n = 8 h). Therefore, the expression for the day-evening-night noise indicator is

$$L_{Rden} = 10 \lg \left[\frac{1}{24} \left(12 \cdot 10^{0.1 L_{Rd}} + 4 \cdot 10^{0.1 (L_{Re} + 5)} + 8 \cdot 10^{0.1 (L_{Rn} + 10)} \right) \right].$$

Location

The physiologically equal amount of noise is most annoying in health and recreation zones intended exclusively for habitation, and least annoying in industrial and infrastructure zones also intended for production and transport. To take account of this sensitivity, urban areas are divided into several—usually four—noise protection zones: For example, noise protection zone I, which includes the most noise sensitive health and recreation areas; noise protection zone II, which includes school and purely residential areas; noise protection zone III, which includes commercial and mixed areas; and noise protection zone IV, which includes the least noise sensitive infrastructure and industrial areas.

Finally, to account for both period of the day and location sensitivity, different *limit values* are specified for every rating level in every noise protection zone. Limit values are regulated by jurisdictions, and these numbers vary considerably from country to country. However, the lowest limit value is always specified for the night rating level in the most sensitive noise protection zone and the highest limit value is always specified for the daytime rating level in the least sensitive noise protection zone. The general

Table 6.2: Good practice limit values for rating levels or noise indicators

Noise Protection	L_{Rden}	L_{Rd}	L_{Re}	L_{Rn}
zone I	$N + 10$	$N + 10$	$N + 5$	N
zone II	$N + 15$	$N + 15$	$N + 10$	$N + 5$
zone III	$N + 20$	$N + 20$	$N + 15$	$N + 10$
zone IV	$N + 25$	$N + 25$	$N + 20$	$N + 15$

idea is presented in Table 6.2, where N, the smallest limit value, is country dependent.

Presence of tones

Tonal noise is characterised by a single frequency component or narrow-band component that emerges audibly from the total noise. The presence of tonal noise increases psychological sensitivity to the overall noise, and is accounted for by increasing measured sound pressure levels by tonal level adjustment K_t. The standard ISO 20065 [50] specifies the engineering method and the standard ISO 1996-2 [26] specifies the simplified, survey method for tone detection. Here we present the survey method.

Noise is determined in one-third octave bands. If a particular band sound pressure level is larger than both adjoining band sound pressure levels for

- 15 dB, for one-third octave bands between 25 Hz and 125 Hz;

- 8 dB, for one-third octave bands between 160 Hz and 400 Hz; or

- 5 dB, for one-third octave bands between 500 Hz and 10 kHz;

the existence of tone is affirmed, and tone level adjustment is applied (Fig. 6.11). The standard recommends tonal level adjustment values between 3 dB and 6 dB [25].

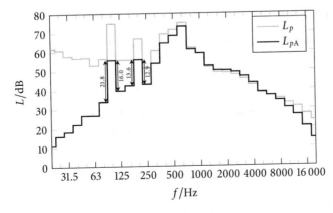

Figure 6.11: Example of tones at 100 Hz and 200 Hz for power electrical trans-former noise in a 50 Hz electric grid. The 100 Hz band sound pressure level is larger than *both* adjacent one-third octave bands for more than the required 15 dB. The 200 Hz band sound pressure level is larger than *both* adjacent bands for more than the required 8 dB [70].

Figure 6.12: The bridge expansion (dilatation) joint is inserted to avoid internal stresses due to expansion of the material (see Section 1.3 on page 6). However, these joints are also an important source of impulsive noise.

Presence of impulsive noise

Impulsive noise is characterised by brief bursts of noise pressure. Examples are pneumatic hammers and vehicles driving over bridge expansion (dilatation) joints (Fig. 6.12). The presence of impulsive noise increases psychological sensitivity to the overall noise and is accounted for by increasing measured sound pressure levels by impulse level adjustment K_i. The standard recommends impulsive level adjustment values of 5 dB for regular impulsive noise and of 12 dB for highly impulsive noise [25]. However, no objective method for impulsive noise detection is specified.

One possible way of objective determination is measuring the noise with F-weighting (fast) and I-weighting (impulse). If the difference of the two obtained results is larger than the prescribed value, the existence of impulsive noise is affirmed, and an impulsive level adjustment is applied.

6.4 Determination of the environmental sound pressure level

Sound pressure levels in the environment can be determined by measurement and by calculation (Fig. 6.13). Each method has its advantages and disadvantages:

Figure 6.13: Two methods for determining the environmental sound pressure level: measurement using a microphone (left) and calculation using a digital terrain model (right) [70].

Measurement	Calculation
higher accuracy at a given moment	lower accuracy at a given moment
noise at time of measurement	long-term averaged noise, neutralising source and atmospheric variations
all noise sources, including unwanted residual noise	all or individual noise sources
one location	wide region
existing noise	existing or planned noise

Most sources have sound power that is strongly period-dependent. Noise quantity due to industry and traffic usually differ between working days and weekends and often has substantial seasonal variation. From the legislation perspective, the relevant value is the year average value, which can be precisely determined only by calculation. Furthermore, often it is important to break the total noise into individual contributions, because various noise sources can have different legislative treatments or because it is necessary to determine the sound source that represents the dominant noise contribution. Therefore, legislation usually prescribe calculation as a relevant method of noise determination.

> **Info box**
>
> Due to its many advantages, calculation is the relevant method for determining sound pressure levels.

However, measurements are still inevitable for nonrepetitive noise incidents (public events, blasts), tonal noise, impulsive noise and peak noise values. Measurements are also used to determine the sound power levels of industrial devices and to verify calculation results.

6.4.1 Measurement of environmental noise

The first method for determining environmental noise, that is, using sound pressure level measurements, is specified in standard ISO 1996-2 [26]. Among other things, the standard defines the requirements concerning instrumentation, operation of the sources, weather conditions and measurement procedures. Frequency-dependent measurements are generally performed in eight octave bands around mid-band frequencies between 63 Hz and 8 kHz.

Even though the measurement devices are able to measure instantaneous noise, the results are averaged for small time periods for convenience. The process is called time weighting and period length is the time constant. Two time weightings, F-weighting (fast) corresponding to time constant 0.125 s and S-weighting (slow) corresponding to time constant 1 s are defined by standards [20]. In addition, I-weighting (impulse) corresponding to time constant 35 ms is used for special applications, such as determining impulse level adjustment.

The impact of residual noise on the measurement is also considered. *Residual noise* is the noise present when the noise source of interest is not present (for example, when a device is turned off). If the sound pressure level of residual noise L_{resid} is at least 10 dB smaller than the sound pressure level of total measured noise L_{meas}, it can be neglected; otherwise, the residual noise should be subtracted in order to get the corrected sound pressure level [26]:

$$L = 10 \lg \left(10^{0.1 L_{meas}} - 10^{0.1 L_{resid}} \right).$$

Further assessment procedures are specified in standard ISO 1996-1 [25]. The most important evaluated sound pressure levels are the following:

- Time-averaged and frequency-weighted sound pressure level, where frequency and time weighting are indicated in the index of the sound pressure level, for example, L_{pAF} for A-weighted F-weighted measurement.

- Maximum time-weighted and frequency-weighted sound pressure level representing the largest level within the stated time interval, for example, L_{AFmax} for A-weighted F-weighted measurement.

- N percent exceedance level representing level that is exceeded by $N(\%)$ of time within the stated time interval, for example, L_{AFN} for A-weighted F-weighted measurement.

- Equivalent continuous sound pressure level representing the frequency weighted continuous measurement over a stated time interval, for example, L_{Aeq} for A-weighted measurement. The calculation expression is

$$L_{AeqT} = 10 \lg \left[\frac{1}{T} \int_T \left(\frac{p_A}{p_0} \right)^2 dt \right],$$

where p_A is the A-weighted instantaneous sound pressure and $p_0 = 20\ \mu Pa$ is the reference sound pressure.

- Rating equivalent continuous sound pressure level, L_{Req}, which was elaborated on in Section 6.3.3.

6.4.2 Calculation of environmental noise

For the second method of determining environmental noise, that is, using sound pressure level calculation, a digital terrain model is created, taking into account the ground surface, sound sources, buildings, noise barriers and other significant objects. Acoustic properties of all objects, including the ground surface, are also taken into account.

The sound pressure level calculation method is specified by standard ISO 9613-2 [32]. First, all finite-dimensional sources are split into point noise sources. Next, each point noise source sound power level L_W is divided into eight octave-band sound power levels L_{Wf} around mid-band frequencies between 63 Hz and 8 kHz.

For a chosen position in space and a particular point noise source, we calculate downwind (in the direction of the wind) octave-band sound pressure level for each of the eight octave bands separately, using expression

$$L_{pf}(\text{DW}) = L_{Wf} + D_C - A,$$

where A is the frequency-dependent octave-band *sound attenuation*:

$$A = A_{\text{div}} + A_{\text{atm}} + A_{\text{gr}} + A_{\text{bar}} + A_{\text{misc}}.$$

Octave-band sound attenuation has the following components:

- A_{div} accounts for geometric divergence due to spherical sound propagation from a point noise source (6.31):

$$A_{\text{div}} = 20 \lg \left(\frac{r}{r_0} \right) + 11 \, \text{dB}.$$

  It is the only sound attenuation component that is frequency independent.

- A_{atm} accounts for sound attenuation due to atmospheric absorption.

- A_{gr} accounts for the noise reflected by the ground surface interfering with noise propagating directly from the source to the receiver.

- A_{bar} accounts for the sound attenuation due to screening obstacles, for example, noise barriers and buildings.

- A_{misc} accounts for sound attenuation due to miscellaneous other effects, for example, foliage, industrial sites and housing.

The equivalent continuous A-weighted downwind sound pressure level for a chosen position in space and a particular point noise source is obtained by summing the octave-band sound pressure levels (6.27) as

$$L_{pA}(\text{DW}) = 10 \lg \left[\sum_f 10^{0.1[L_{pf}(\text{DW}) + A_f]} \right],$$

where A_f denotes standard A-weighting (Fig. 6.10 on page 203).

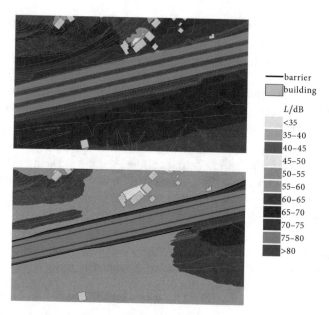

Figure 6.14: Noise map in the vicinity of the highway from the digital terrain model in Fig. 6.13, 2 m above the ground. Top is the situation without noise abatements, and bottom is the situation with noise abatements. See Fig. 6.18 on page 215 for the cross-sectional representation of the same area. The colour scheme used is defined by standard DIN 18005-2 [70].

Finally, the long-term average A-weighted sound pressure level is obtained from

$$L_{pA}(\text{LT}) = L_{pA}(\text{DW}) - C_{\text{met}},$$

where C_{met} is the meteorological correction.

A more precise method of calculating the sound attenuation of noise as a result of atmospheric absorption is specified in standard ISO 9613-1 [31].

The year averaged sound pressure levels are usually calculated

- for a given region and given sound sources,
- for all periods of day,
- for exposed façades of all buildings and
- at a constant height above ground, usually 2 m or 4 m.

Sound pressure level distribution is represented graphically in the form of a noise map (Fig. 6.14). The region is usually classified using 5 dB steps and coloured according to various national colour schemes.

6.4.3 Road traffic

Road traffic is the single largest contributor of noise in urban areas (Table 6.3). For this reason, we will loosely describe a calculation method for its sound power level.

Table 6.3: Estimated number of people in the European Union exposed to noise levels $L_{den} \geq 55\,dB$ [80].

Noise Source	Exposed People
road traffic	125 000 000
rail traffic	8 000 000
aircraft traffic	3 000 000
industry	300 000

The European commission directive 2015/996 [61] established common noise assessment methods in the European Union, usually referred to by the acronym CNOSSOS-EU (common **no**ise assessment method**s**). With these methods the traffic noise source of roads or road lanes is represented by linear noise sources 0.05 m above the road surface. Their linear sound power levels depend on several parameters:

- *Vehicle type*. Vehicles are divided into four categories: light motor vehicles, medium heavy vehicles, heavy vehicles and powered two-wheelers, the latter being further subdivided into two subcategories.

- *Average velocity*. The average velocity is required for each vehicle category.

- *Traffic volume*. The number of vehicles per hour is required for each vehicle category.

- *Road gradient*. The road gradient affects both engine load and engine speed through the choice of gear and therefore the propulsion noise emission of the vehicle. The effect depends on the slope.

- *Traffic flow type*. Acceleration and deceleration before and after crossings with traffic lights and roundabouts affect the rolling and propulsion noise. The effect depends on distance to the nearest intersection.

- *Road surface*. Porous road surfaces generate less rolling noise and absorb more propulsion noise. Velocity dependent corrections are defined for fourteen different types of road surfaces in relation to the reference road surface.

- *Air temperature*. The rolling sound power level slightly decreases as the air temperature increases.

- *Studded tyres*. If a significant number of light vehicles use studded tyres for several months of the year, the induced velocity dependent effect on rolling noise is taken into account.

The CNOSSOS-EU methods specify the calculation of the linear sound power level (6.33) of a road with one vehicle per hour. The sound power level depends on the velocity and the vehicle category. Fig. 6.15 shows the linear sound power level for the first three vehicle categories. Note that in most practical situations (higher velocities, light motor vehicles), the linear sound power level of the road increases with velocity.

The noise emission is the result of three contributions:

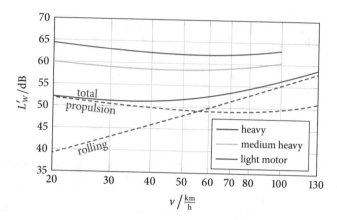

Figure 6.15: Linear sound power levels of a road with one vehicle per hour for first three vehicle categories, according to the CNOSSOS-EU methods. The contribution of the rolling and the propulsion is shown for the light motor vehicles.

1. *Propulsion noise* (engine and exhaust system) is the dominant contribution to the sound power at lower velocities; however, this contribution is constantly being reduced by new car designs.

2. *Rolling noise* (interaction between road surfaces and rubber tires) is the dominant contribution at higher velocities. This contribution can be influenced by both tire design and surface pavement.

3. *Aerodynamic noise* increases with vehicle velocity. Note that in CNOSSOS-EU aerodynamic noise is incorporated in the rolling noise source (Fig. 6.15).

The traffic volumes for all vehicle (sub)categories must be determined by automatic traffic counters or traffic demand models. If n_1, n_2, n_3, n_{4a} and n_{4b} are the number of vehicles per hour for all vehicle (sub)categories, whereas L'_{W1}, L'_{W2}, L'_{W3}, L'_{W4a} and L'_{W4b} are the linear sound power levels of one vehicle per hour for all vehicle (sub)categories, the total linear sound power level is (6.27):

$$L'_W = 10\lg\left(n_1\,10^{0.1L'_{W1}} + n_2\,10^{0.1L'_{W2}} + n_3\,10^{0.1L'_{W3}} + n_{4a}\,10^{0.1L'_{W4a}} + n_{4b}\,10^{0.1L'_{W4b}}\right).$$

The CNOSSOS-EU methods [61] also established assessment methods for railway noise, industrial noise, aircraft noise and noise propagation. For example, the railway traffic noise source is represented by two linear noise sources 0.5 m and 4.0 m above the tracks.

6.5 Noise abatement

A wide range of mitigation methods can be used to reduce noise levels and negative influences on the population. They can be roughly divided into the following four categories:

1. *Reducing the sound power of noise sources* is the most effective method. It strongly depends on the noise source type. For illustration purposes, we will discuss methods of reducing sound power levels for *road and rail traffic* (see also Section 6.4.3):

    a) *Decreasing vehicle velocity* is the simplest method, achieved by imposing lower speed limits, or, for road traffic, building roundabouts.

    b) *Quieter vehicle propulsion* due to technological advancements and ever stricter legislation is another efficient method. The most effective step in this direction is the gradual replacement of internal combustion engines with hybrid or electric ones. For rail traffic, electrification of rail lines is always accompanied by the replacement of diesel-powered locomotives with electric-powered locomotives and immediate reduction of sound power levels.

    c) *Quieter braking system* is especially efficient for rail traffic. The sound power level in proximity to stations can be reduced by using electric braking systems instead of air braking systems.

    d) *Porous substratum* reduces sound reflection and thus effectively reduces directivity correction (Section 6.2.4). Porous road surfaces and rail lines with track ballast have significantly smaller emissions compared to smooth asphalt and tracks mounted directly to the continuous slab of concrete, respectively. On the other hand, porous road surfaces also diminish sound power levels due to improved interaction between the tires and the surface.

    e) Many other methods are available. For example, a significant reduction in rail traffic emissions can be achieved by using of timber sleepers instead of concrete sleepers and by not attaching rail tracks directly to bridge structures and thus avoiding the transition of vibrations.

2. *Active measures* or noise barriers are aimed at preventing sound propagation into the environment. Noise barriers may be designed as a wall construction, earthwork or combination of both (for example, a wall atop an earth berm). The effectiveness is governed by their physical dimensions (especially height). In the most extreme cases, the entire noise source (road, railway) can be surrounded by a structure, or dug into a tunnel using the cut-and-cover method.

    Info box

    Active measures prevent sound propagation into the environment, whereas passive measures prevent propagation into buildings.

    Acoustic shadow, the area into which sound waves fail to propagate, forms behind the barrier (Fig. 6.16). It should be noted that the acoustic shadow is increased if the barrier of the same dimensions approaches the noise source. In general, the efficiency of the barrier can be strongly influenced by proper positioning. Further, in order to satisfactorily prevent direct noise propagation, wall barriers must have a large airborne sound reduction index (see Section 7.1 on page 217) without air gaps that represent acoustic bridges (see Section 7.4 on page 229).

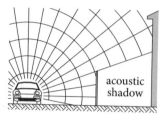
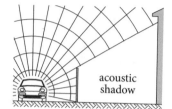

Figure 6.16: Acoustic shadow behind a noise barrier. Note that the noise barrier closer to the noise source (right) creates a larger acoustic shadow than the noise barrier of the same height farther from the source (left).

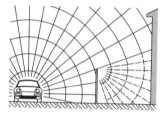
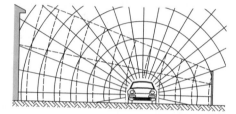

Figure 6.17: Diffraction causes indirect propagation of noise behind the noise barrier (dashed lines, left). Reflection increases the noise on the opposite side of the noise source (dashed lines, right).

Despite the fact that noise fails to propagate behind the noise barrier directly, there is still a small, but nonnegligible presence of noise due to *sound diffraction*, especially at low frequencies (Fig. 6.17, left). Diffraction is taken into account by standard ISO 9613-2 [32].

Finally, *sound reflection* from the barrier increases the noise on the opposite side of the noise source (Fig. 6.17, right). In order to prevent this effect, barrier absorbance (see Section 7.1 on page 217) should be as large as possible.

3. *Passive measures* are used to mitigate noise in situations where reducing sound power and active measures are not effective enough. These measures are intended to protect the interior of the buildings from the noise by increasing the airborne sound reduction index of the external building elements. Because the most noise-permissive building components are windows, the most effective and most common measure is to use windows with a large airborne sound reduction index.

4. *Urban planing* can be used to separate noise sources (for example, industrial zones and transport corridors) from areas sensitive to noise (for example, residential zones). Good zone planing can substantially reduce or even eliminate the need for other noise abatement measures.

The effect of three of these abatement methods is visualised in Fig. 6.18. Abatement includes one sound power reduction measure, a porous road surface (A). There are also three active measures: earth berm of height 1 m (B_1), noise barrier of height 6 m (B_2) and noise barrier of height 4 m (B_3). Finally, because the sound pressure levels for higher

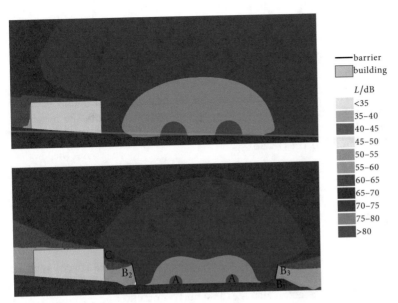

Figure 6.18: The effect of three abatement methods. The sound pressure levels for the cross-section in the vicinity of the highway from the digital terrain model in Fig. 6.13 on page 207 are presented, without (top) and with (bottom) noise abatements. Abatement consists of one sound power reduction measure (A), three active measures (B₁, B₂, B₃) and one passive measure (C). See Fig. 6.14 on page 210 for the situational representation of the same area. The colour scheme used is defined by standard DIN 18005-2 [70].

floors of the residential building are still too large, the passive measure in terms of increasing the airborne sound reduction index (C) is applied there.

Problems

6.1 The sound power of an isotropic point source is 2.0×10^{-6} W. Calculate the sound power level as well as the intensity and sound pressure level 10 m from the source. ($63\,\mathrm{dB}$, $1.6 \times 10^{-9}\,\mathrm{W/m^2}$, $32\,\mathrm{dB}$)

6.2 The sound pressure level is 70 dB at distance 3.0 m from the isotropic linear sound source. Determine the distance from the source at which the sound pressure level is 58 dB. ($48\,\mathrm{m}$)

6.3 Three individual isotropic point sound sources of sound power levels 55 dB, 50 dB and 45 dB are positioned closely to each other. What is the overall sound power level? What is the smallest distance from these sources for which they are no longer audible? ($56.5\,\mathrm{dB}$, $189\,\mathrm{m}$)

6.4 The sound measurement location is 15 m from an electric motor of sound power level 90 dB and 25 m from the road of linear sound power level 75 dB. What is the measured sound pressure level? (57.4 dB)

6.5 When two sound sources operate, the measured sound pressure level is 80 dB. When the first sound source is turned off, the measured sound pressure level is 75 dB. What is the sound pressure level due to the first sound source? (78.3 dB)

6.6 The sound pressure level between 8:00 and 11:00 is 60 dB and between 11:00 and 18:00 is 50 dB. What is the average sound pressure level? (55.7 dB)

6.7 The sound pressure level gain due to physiological effects, such as the shape of the head and the outer ear, is 20 dB [69]. For the quietest sound still detectable, calculate the sound power entering the human body through the tympanic membrane, if the effective area of the tympanic membrane is 43 mm^2. (4.3×10^{-15} W)

7 Building acoustics

In a majority of the previous chapter, sound was propagated under *free sound field* conditions. In this chapter, we are primarily interested in closed spaces, where the area of reflecting surfaces is so large that reflected sound waves predominate the sound waves coming directly from the sound sources. This profoundly changes the sound model of the previous chapter and requires a different study approach.

7.1 Acoustic properties of materials

We have already pointed out that sound can propagate as long as material particles are present in all states of matter. In air, sound can propagate almost unhindered, apart from the small sound attenuation due to atmospheric absorption. In enclosed spaces and in the presence of solid matter objects, sound reflections and dissipation become relevant.

When a sound wave strikes a solid object, part of the incident sound is absorbed, part is transmitted and part is dissipated (Fig. 7.1). If we denote sound intensity for incident sound I, for reflected sound I_ρ, for dissipated sound I_δ and for translated sound I_τ, we can define *reflectance* ρ,

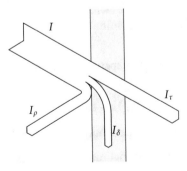

Figure 7.1: Part of the incident sound wave (I) is reflected (I_ρ), part is transmitted (I_τ) and part is dissipated (I_δ).

dissipance δ and *transmittance τ* as

$$\rho(f) = \frac{I_\rho(f)}{I(f)}, \tag{7.1}$$

$$\delta(f) = \frac{I_\delta(f)}{I(f)}, \tag{7.2}$$

$$\tau(f) = \frac{I_\tau(f)}{I(f)}. \tag{7.3}$$

Here we took into account that all three physical quantities depend on the sound frequency. Note that their value range is $0 \le \rho(\lambda), \delta(\lambda), \tau(\lambda) \le 1$.

Because energy is conserved, the sum of the sound intensities for reflected, dissipated and translated sound waves should be equal to the sound intensity for the incident sound wave as

$$I_\rho(f) + I_\delta(f) + I_\tau(f) = I(f),$$

which together with the definition of reflectance, dissipance and transmittance, leads to

$$\rho(f) + \delta(f) + \tau(f) = 1. \tag{7.4}$$

In building acoustics, the most important physical quantity is *absorbance* α. Absorbance is the ratio of absorbed sound intensity, which is the sum of transmitted *and* dissipated sound intensity, to incident sound intensity:

$$\alpha(f) = \frac{I_\alpha(f)}{I(f)} = \frac{I_\delta(f) + I_\tau(f)}{I(f)} \tag{7.5}$$

$$\implies \alpha(f) = \delta(f) + \tau(f). \tag{7.6}$$

From the energy conservation expression (7.4), we get

$$\rho(f) + \alpha(f) = 1.$$

Note that *absorptance* (2.36), defined for radiation, and *absorbance* (7.5) have completely different meanings and should not be confused with each other. Absorptance (2.36) agrees with dissipance (7.2). Both reflectances, (2.35) and (7.1), and both transmittances, (2.37) and (7.3), have the same meanings.

Typically, transmittance values span several orders of magnitude, and it is more convenient to use *sound reduction index R*, which is defined as

$$R(f) = -10 \lg \tau(f). \tag{7.7}$$

The minus sign ensures that the sound reduction index is positive.

By taking the logarithm of the definition for transmittance (7.3) as

$$I_\tau = \tau I,$$

$$10 \lg \left(\frac{I_\tau}{I_0} \right) = 10 \lg \tau + 10 \lg \left(\frac{I}{I_0} \right),$$

we can conclude that the sound reduction index is simply the difference of incident L_p and transmitted sound pressure level $L_{p\tau}$:

$$R(f) = L_p(f) - L_{p\tau}(f). \tag{7.8}$$

This equation is very useful for sound barriers in *free sound field* situations.

Similarly, the difference between incident L_p and reflected sound pressure level $L_{p\rho}$ can be written as

$$\Delta L(f) = L_p(f) - L_{p\rho}(f) = -10\lg\rho(f). \tag{7.9}$$

The equation is indispensable for the investigation of sound reflections.

7.2 Room acoustics

Closed spaces lead to abundant waves reflections, which create new phenomena that are not present or not important in free sound field conditions.

7.2.1 Echo

The first important phenomenon is *echo*, which refers to a reflected sound wave. The sound signal from the source to the receiver can take either a direct path or an echo path, and because the latter path is longer, there is a *time delay* between the arrival of the direct sound and the reflected sound (Fig. 7.2). The time delay can be calculated using speed of sound c and length of both paths d_1 and d_2 as

$$\Delta t = \frac{d_2}{c} - \frac{d_1}{c} = \frac{d_2 - d_1}{c}.$$

For time delays smaller than 100 ms for music and 50 ms for speech, both sound waves are fused into one in our brains and we do not perceive an echo. This phenomenon is called the *Haas effect*. Because a time delay of 50 ms corresponds to the path length difference of 15 m, time delay is an important issue primarily in larger rooms.

The presence of wave reflections can also create *standing sound waves*, which create areas within the room where sound energy is diminished or enhanced (Fig. 5.12 on page 182). We avert the creation of standing waves by evading parallel walls within the room.

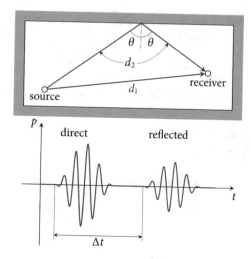

Figure 7.2: Direct and reflected sound (echo). Because the echo path d_2 is larger than the direct path d_1, there is a time delay Δt between the arrival of practically the same sound signals. Note that the reflected sound has a smaller intensity due to the longer path and absorbance.

7.2.2 Diffuse sound field

A large area of reflecting surfaces induce repeated reflections and diffractions of sound. Although the sound pressure level of one reflected wave is always smaller than the sound pressure level of the direct wave, reflected waves predominate due to their quantity. This creates a relatively uniform distribution of sound energy within a room and the relatively constant sound pressure level, which is identified as *diffuse (reverberant) sound field* conditions. More precisely, a diffuse sound field is a region with

- a constant sound pressure value over the whole region, and

- no preference of direction regarding where the sound is coming from for an arbitrary point within the region.

However, in close sound source proximity, direct sound waves predominate and conditions resemble those for free sound field. This limited region is identified with *direct sound field* conditions (Fig. 7.3).

The value of the sound pressure level of the diffuse sound field depends on two processes:

1. Sound sources in the room add sound energy. They can be either sound sources within the room or external sound sources. Note that for external sound sources, we can assume that building elements through which sound enters the room represent the surface sound sources.

2. Sound absorption subtracts sound energy. Sound energy is primarily absorbed by surfaces within the room. Note that absorption here includes sound that is transmitted out of the room. Some sound energy is also absorbed by the air, but this effect is usually negligible.

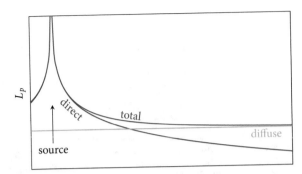

Figure 7.3: Direct and diffuse sound fields for a point sound source. Close to the source, direct sound waves prevail, creating a direct sound field region. Far from the source, reflected sound waves prevail, creating a diffuse sound field region.

If added sound energy is larger than subtracted sound energy, the total sound energy within room E increases. Conversely, if added sound energy is smaller than subtracted sound energy, the total sound energy within room E decreases. If added and subtracted sound energy are the same, we get a *stationary* situation and the total sound energy in the room is constant. All three situations can be described by the differential diffuse sound field equation of

$$\frac{dE}{dt} = P - P_\alpha, \tag{7.10}$$

where P is the aggregate sound power of all sound sources, and P_α is the absorbed sound power.

We will first elaborate on the absorbed sound power by a small surface of area S and absorbance α. Assuming that all sound waves strike the surface perpendicularly, incident sound power P_{inc} is (6.17)

$$P_{inc} = S I.$$

Under diffuse sound field conditions, plane waves come uniformly from all directions. This means the actual sound intensity of the incident sound waves will only amount to half of the total sound intensity because only waves going towards the surface (and without those going away from the surface) are considered. Furthermore, because a particular plane wave comes under incident angle θ_i, the effective interception surface area S' is (Fig. 7.4)

$$S' = S \cos \theta_i,$$

and the intercepted power is reduced by term $\cos \theta_i$:

$$P_{inc}(\theta_i) = \frac{1}{2} S I \cos \theta_i.$$

Using equation (7.5), we can finally calculate the absorbed energy of the particular plane wave by surface of area S:

$$P_{\alpha,S}(\theta_i) = \frac{1}{2} \cos \theta_i \, \alpha \, S I.$$

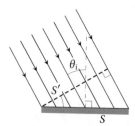

Figure 7.4: Intercepted sound energy for a surface of area S and angle of incidence θ_i. When the sound wave does not strike the surface perpendicularly, an element of surface intercepts smaller sound power for the same sound intensity.

In order to get the absorbed sound power of all plane waves within the diffuse sound field, we have to integrate over the half-space (Fig. 7.5). All plane waves with angle of incidence θ_i come from a solid angle (8.16) that amounts to

$$d\Omega = 2\pi x\, ds = 2\pi \sin \theta_i\, d\theta_i,$$

so the total absorbed sound power by area S is

$$P_{\alpha,S} = \int P_{\alpha,S}(\theta_i)\, d\Omega = \frac{1}{2}\,\alpha\, S I \int_0^{\pi/2} 2\pi \sin \theta_i \cos \theta_i\, d\theta_i = \frac{1}{4}\,\alpha S\, I.$$

In order to calculate the total absorbed sound power by all objects in the room, we have to sum over all surfaces in the room, including the room wall surfaces as

$$P_\alpha = \frac{1}{4} I \sum_i \alpha_i S_i = \frac{1}{4} I A, \tag{7.11}$$

where

$$A = \sum_i \alpha_i S_i \tag{7.12}$$

is the sum of the products of the room surface areas and their absorbances, which is called *equivalent absorption area A (m^2)*.

> **Info box**
>
> Equivalent absorption area is the most important acoustic property of the room.

Next, we calculate the total sound energy within the room, assuming that diffuse sound field conditions (constant sound intensity) essentially extend

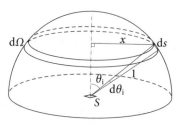

Figure 7.5: A solid angle of plane waves with angle of incidence θ_i above the small surface.

to the total room volume. Using equations (6.18) and (6.21), we get

$$E = e\, V = \frac{I\, V}{c},$$ (7.13)

where V is the room volume.

Using equations (7.11) and (7.13), we can transform the diffuse sound field equation (7.10) in the form

$$\frac{V}{c}\frac{dI}{dt} = P - \frac{1}{4}I\, A.$$ (7.14)

We have obtained the differential equation for the sound intensity in the diffuse sound field conditions.

Using equation (7.14), we can calculate the diffuse sound field in the *stationary* situation. Because the added and absorbed sound energy are the same, the sound intensity does not change, and we get

$$I = \frac{4\, P}{A}.$$

The corresponding sound pressure level is (6.24)

$$L_p = 10\lg\frac{I}{I_0} = 10\lg\left(\frac{4P}{A}\frac{S_0}{P_0}\right) = 10\lg\left(\frac{P}{P_0}\frac{4S_0}{A}\right),$$

$$L_p = L_W - 10\lg\frac{A}{4S_0},$$ (7.15)

where $S_0 = 1\,\mathrm{m}^2$. Compare that expression with the one for free field conditions (6.31).

7.2.3 Reverberation time

Another interesting situation occurs when all sound sources are turned off, that is, $P = 0$. The diffuse sound field equation (7.14) then transforms to

$$\frac{dI}{I} = -\frac{c\, A}{4\, V}dt = -\frac{dt}{t_0},$$

where

$$t_0 = \frac{4\, V}{c\, A}$$ (7.16)

is the lifetime of sound intensity decay. Using the boundary condition that at time $t = 0$ sound intensity is equal to $I(0)$, the solution of the equation is

$$I = I(0)\, e^{-t/t_0}.$$ (7.17)

Putting that into (6.24), we get

$$L_p = L_p(0) + 10\lg\left(e^{-t/t_0}\right) = L_p(0) - \frac{10}{\ln(10)}\frac{t}{t_0}.$$

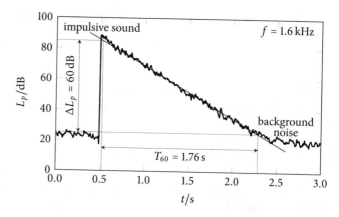

Figure 7.6: Reverberation time T_{60} is defined as a time in which the sound pressure level of the diffuse sound field reduces by 60 dB. Reverberation time is determined by generating a sound pulse and measuring the sound pressure level as the noise drops back to its background value. Figure shows the result of the measurement of reverberation time in a community hall, by bursting a balloon [70].

On turning sound sources off, the sound pressure level *linearly decreases* in time due to sound reflections of the existing sound waves. It is common to define a *reverberation time* T_{60} as a time in which sound intensity decreases for a factor of 10^6, which corresponds to the sound pressure decrease of factor 10^3, and sound pressure level decrease of $10 \lg 10^6 = 60$ dB.

Reverberation time is frequency dependent. It is measured by generating a sound pulse and measuring the sound pressure level as the noise drops back to its background value (Fig. 7.6).

We can theoretically calculate reverberation time using (7.17):

$$\frac{I}{I(0)} = -e^{-T_{60}/t_0} = 10^{-6}$$

$$\implies T_{60} = t_0 \ln(10^6) = \frac{4 \ln(10^6)}{c} \frac{V}{A}.$$

The obtained expression, which connects the reverberation time with equivalent absorption area A (7.12) and room volume V, is called the *Sabine equation* and is usually expressed in the form

Info box

Measurements of reverberation time are often used to indirectly determine the equivalent absorption area.

$$T_{60} = \left(0.163 \, \frac{s}{m}\right) \cdot \frac{V}{A}. \tag{7.18}$$

The Sabine formula allows a quick determination of the equivalent absorption area. The reverberation time is determined with a microphone as in Fig. 7.6, after which the equivalent absorption area is extracted from (7.18).

Reverberation time is the good indicator of room acoustics. A short reverberation time gives the impression of 'dead' sound, that is, the loss of sound richness, whereas a long reverberation time gives the impression of 'muddy' sound, that is, the loss of sound articulation. As a rule of thumb, different frequency averaged reverberation times are appropriate for different room purposes:

- lecture halls: $0.7\,s < T_{60} < 1.2\,s$
- music halls: $1.2\,s < T_{60} < 2.5\,s$
- churches: $2.5\,s < T_{60}$

The Sabine equation instructs us that we can tweak the reverberation time by changing the absorbance of the surfaces in the room.

7.3 Sound absorbers

In the previous sections we discussed the relation between sound reflections and the room purpose. In order to obtain a quality sound environment, it is important to control time delay and reverberation time by adjusting surface absorbances. Because the absorbance of hard smooth surfaces is practically frequency independent and rather small ($\alpha \approx 0.05$ for concrete, $\alpha \approx 0.10$ for timber), reflections are boosted by keeping wall surfaces uncovered.

On the other hand, reducing reflections and increasing absorbance is a more complex task. Obviously, absorbance is increased either by increasing dissipance or transmittance (7.6). However, increasing wall transmittance works both ways and boosts unwanted penetration of the external noise. We are therefore compelled to increase wall dissipance only by attaching materials or devices with high dissipance. Those materials and devices are called sound absorbers, and their frequency-dependent absorbance is determined in accordance with standard ISO 354 [22].

Dissipance is actually transformation of sound energy into internal energy. Sound energy is equal to kinetic energy of fluid particles and is transferred by elastic collisions between those particles. The important issue is the mean free path of air (the average distance a fluid molecule can move before colliding with another), which is about 68 nm [71] for air at atmospheric pressure. However, due to the friction between the fluid and the solid surfaces, as well as fluid viscosity, some of the kinetic energy is transformed into the internal energy of the fluid and adjacent solid material. This transfer is strongest in a thin air layer near solid surfaces as the velocity of fluid particles next to the surface is forced to zero (see also Section 2.3.1 on page 49). Furthermore, the impact of the sound wave on a solid material dissipates some sound energy by flexing the solid frame. Both mechanisms account for small but nonetheless important absorbance that exists even with hard smooth surfaces.

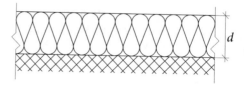

Figure 7.7: Porous sound absorber is usually mineral wool of thickness d mounted on the surface of the solid wall.

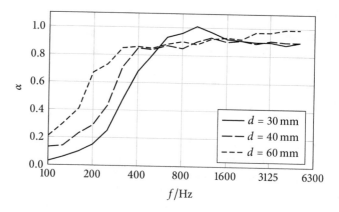

Figure 7.8: Absorbance of mineral wool for three different porous layer thicknesses. Absorbance at low frequencies increases with the thickness.

7.3.1 Porous sound absorbers

The most obvious increase of dissipation is achieved by increasing the surface area. If the solid surface is highly porous, a substantial portion of the sound wave penetrates the material. The sound wave eventually reflects from the porous surface, but before returning to fluid bulk, there is a high probability that the sound wave will encounter another porous surface and reflect again. Many subsequent reflections can greatly improve dissipation. This scattering is essential to the performance of *porous sound absorbers*.

It is important to note that not all highly porous materials are suitable for sound absorbers. Firstly, sound waves cannot penetrate materials with closed-cell structures, such as EPS, XPS or aerogel. Secondly, even in materials with open-cell structures, if the characteristic dimension of the pores (mean distance between pore walls) is comparable or smaller than the mean free path of air, such as concrete and gypsum plaster, sound waves will not be able to efficiently penetrate the material. Only materials with open-cell structures and large pores, such as mineral wool or carpet, allow efficient penetration of sound waves.

Therefore, the porous sound absorber is essentially an open-cell large pore material, usually mineral wool, mounted on the surface of the solid wall (Fig. 7.7). Because porous material is fragile and may collect dust, it is often covered by fabric, thin plastic film or metal sheeting.

However, porous sound absorber absorbance is dependent on its thickness d and sound wave frequency (Fig. 7.8). Because of the practically

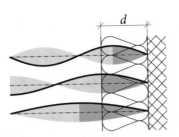

Figure 7.9: Sound waves of three different wavelengths striking the solid wall with the porous layer. Absorbance for the top and middle sound waves will be large as the quarter-wavelength is smaller than the absorber thickness, $\lambda/4 \leq d$. Absorbance for the bottom sound wave will be small as its quarter-wavelength is larger than the absorber thickness, $\lambda/4 > d$.

total reflectance from the solid wall, sound waves form standing waves with a node on the surface of the solid wall (see Section 5.5 on page 179), as shown in Fig. 7.9. Note that positions of the largest sound pressure and consequently the largest particle energy (velocity) are at $\lambda/4, 3\lambda/4, 5\lambda/4...$ from the solid wall. On the other hand, a large *flow resistance* that leads to dissipative interaction appears only within the porous layer. Therefore, the absorbance of the porous sound absorber is very efficient only if the location of the largest sound pressure coincides with the porous layer, that is

$$\alpha \approx \delta \approx 1 \iff \frac{\lambda}{4} \leq d. \tag{7.19}$$

From (5.10), we conclude that porous sound absorbers are most effective at high frequencies and can be improved by increasing the porous layer thickness (Fig. 7.8). However, making porous material very thick takes too much valuable room space, so for lower frequencies other types of sound absorbers based on the principle of *resonance* are used instead.

7.3.2 Membrane sound absorbers

The other type of sound absorber, called a membrane sound absorber, is designed as a panel attached on the solid wall with spacers (Fig. 7.10). The system panel plus air trapped behind the panel represent an oscillator whose resonant frequency f_0 depends on the thickness of air layer d and surface density of the panel ρ_A. The approximate equation is

$$f_0 = \frac{c}{2\pi} \sqrt{\frac{\rho}{\rho_A d}},$$

where ρ is the density of air, and c is the speed of sound.

The absorption process consists of two steps:

1. When the frequency of the sound wave that strikes the panel approximately matches the resonant frequency, the system starts oscillating, which means that sound energy of open air is transformed to the oscillation energy of the membrane sound absorber.

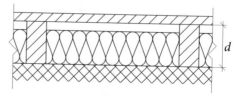

Figure 7.10: Membrane sound absorber designed as a panel is attached on the solid wall with spacers. Empty space is partially filled with porous material to increase dissipation.

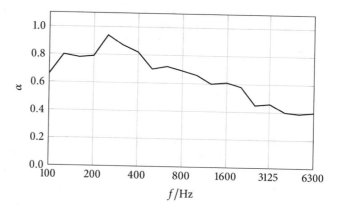

Figure 7.11: Absorbance of the membrane sound absorber with a resonance frequency of 250 Hz.

2. Oscillation energy is converted to internal energy either due to friction of the panel structure or due to viscosity of the air trapped behind the panel. The dissipation in the air can be significantly increased by increasing the flow resistance by filling the empty space with the porous material.

Obviously, the absorbance is the largest for frequencies near the resonance frequency (Fig. 7.11). This frequency can be conveniently tuned to lower sound frequencies, which are not well absorbed by porous sound absorbers.

7.3.3 Helmholtz sound absorbers

Another form of sound absorber is based on the resonance of the air trapped in a cavity. The phenomenon is called a Helmholtz resonance and can be observed when blowing across the top of an empty bottle. The absorption process consists of two steps:

1. When the frequency of the sound wave that enters the cavity approximately matches the resonant frequency, the air within the cavity starts oscillating, which means that the sound energy of open air is transformed to the oscillation energy of the cavity air.

2. Oscillation energy is converted to internal energy due to the viscosity of the air within the cavity. The dissipation can be significantly in-

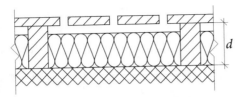

Figure 7.12: Helmholtz sound absorber is usually designed as a perforated panel attached on the solid wall with spacers. The empty space is partially filled with porous material to increase dissipation.

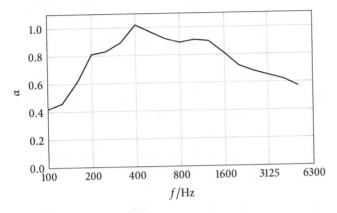

Figure 7.13: Absorbance of a perforated panel absorber. Because the effects of the membrane and Helmholtz resonance are combined, we obtain a wider frequency band with large absorbance.

creased by increasing flow resistance through filling the empty space with the porous material.

The construction of this type of absorber usually consists of a perforated panel above an airspace, as shown in Fig. 7.12. The resonant frequency depends on the geometry of the absorber (size of holes, their spacing, the thickness of airspace and panel).

Note that the structure on Fig. 7.12 can actually combine the effects of both the membrane and Helmholtz resonance. Obviously, the absorbance is largest for frequencies near both resonance frequencies. By conveniently tuning them, a wider frequency band with a large absorbance can be created (Fig. 7.13).

7.4 Sound transmission and insulation

In general, there are three methods of sound transmission within a building (Fig. 7.14):

1. *Airborne noise transmission.* A sound source (for example, speech, loudspeakers, car) creates air sound waves in a room or in an environment, which are then transmitted by the separating building ele-

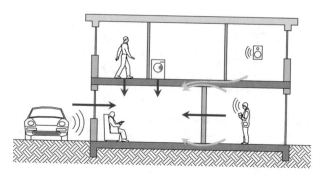

Figure 7.14: Means of sound transmission within a building: airborne transmission (blue lines), impact transmission (red lines) and flanking transmission (yellow lines).

ment into the adjacent room. Transmission is facilitated by means of building element vibrations that are caused by air sound waves.

2. *Impact noise transmission.* Building element vibrations are created directly by the collision with an object (for example, walking and vibrating machines in contact with the floor). Vibrations create sound in the adjacent room (most notably below).

3. *Flanking noise transmission.* Sound is transmitted by flanking building elements, bypassing the separating building element above or below. This method can facilitate sound transmission even between rooms that are not adjacent.

Another important issue is the *acoustic bridge*. Acoustic bridges are parts of building element separating two environments that transfer considerably larger amounts of sound energy than their surroundings, or installation items that provide alternative paths for the sound and vibration transmissions between different parts of the buildings. Typical examples include the following:

- *Cracks and gaps.* Even the smallest air gap in the building element can significantly increase the sound transmission. All cracks and gaps should be properly sealed, in particular in and around window frames. The same applies for cracks and gaps in sound barriers (see Section 6.5 on page 212).

- *Air ducts and installation.* Air duct channels should not connect different rooms, both vertically and horizontally.

- *Water and heating installations.* Pipes, especially low-mass plastic pipes, should be carefully wrapped with insulation. Note that sometimes noise is produced by the installation itself (for example, water hammer).

- *Joints in the masonry.* The effect of these acoustic bridges is reduced by plastering the wall on both sides.

- *Installations in the insulation layer.* By interrupting the sound insulation layer, for example, between the floor slab and the supporting structure, vibrations and sound can easily cross.

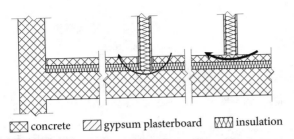

⊠ concrete ⊿ gypsum plasterboard ▧ insulation

Figure 7.15: Methods for decreasing impact and flanking sound transmission. Floating floor method (left) decreases impact sound transmission by preventing transmission of vibrations from the floor slab to the building construction. Partition walls that are mounted directly on the construction (centre) allow far less flanking sound transmission as compared to partition walls mounted to the floor slab (right).

- *Electrical sockets.* They increase sound transmission by interrupting both the wall and sound insulation layer, especially when two electrical sockets are placed on the opposite sides of the same wall.

Note that flanking noise transmission and sound transmission through acoustic bridges significantly reduces the sound reduction index of the building elements. Because neither of them can be completely removed, the effective noise reduction index of the wall in the constructed dwelling is always smaller than the theoretical noise reduction index. This fact should be taken into consideration in the building design.

Several building details can considerably improve sound insulation:

- In drywall, the space between two gypsum plasterboards should be filled with the insulation layer, which attenuates sound propagation.

- The sound reduction index can be further increased by repetitive switching between gypsum plasterboards and insulation layers. Each change from gypsum plasterboard to insulation layer and vice versa reflects part of the sound.

- *Floating floor* is the construction method in which the floor slab (or floor assembly) is not directly mounted to the supporting structure but completely separated from it by underlayment. Underlayment impairs the transmission of vibrations thus reducing the impact sound transmission. On the left side of Fig. 7.15, concrete slab (screed) is laid on precompressed insulation (for example, polystyrene). Note that for the same reason, floor slabs must be on the edges separated from the walls.

- *Partition walls* should be first mounted directly to the building construction, whereas floor slabs should be laid afterwards (Fig. 7.15, centre). If a partition wall is mounted on the floor slab, flanking sound transmission is significantly increased due to the shortened sound transmission path and vibration transmission through the slab.

- *Dropped ceilings* also reduce flanking sound transmission by elongating the sound transmission path (Fig. 7.16).

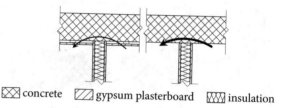

$\boxtimes$ concrete $\diagdown\!\!\diagdown$ gypsum plasterboard $\bowtie$ insulation

Figure 7.16: A dropped ceiling (left) allows less flanking sound transmission compared to a bare ceiling (right).

7.4.1 Mass law

Airborne sound transmission through real building elements is a very complex process. In order to get the basic theoretical description, we will assume that the element is a homogeneous and flexible panel and neglect the sound dissipation within it (Fig. 7.17). Pressure wave functions for incident p_ι, reflected p_ρ and transmitted p_τ waves are

$$p_\iota = A_\iota \sin(\omega t - kz),$$
$$p_\rho = A_\rho \sin(\omega t + kz + \varphi_\rho),$$
$$p_\tau = A_\tau \sin(\omega t - kz + \varphi_\tau),$$

where the '+' sign for reflected wave indicates that it is moving backward (5.8). Note that, in general, all three sound waves have different phases, so we use φ_ρ and φ_τ to describe the phase difference of the reflected and transmitted waves with respect to the incident sound wave, respectively. Relations between velocity and pressure for the preceding wave functions are (6.7)

$$\frac{p_\iota}{\rho c} = v_\iota, \quad \frac{p_\rho}{\rho c} = -v_\rho, \quad \frac{p_\tau}{\rho c} = v_\tau.$$

We have to determine four unknown constants. We start by noting that panel velocity v_{pan} and air velocities on both panel surfaces must be the same:

$$v_\iota(z=0) + v_\rho(z=0) = v_{\text{pan}} = v_\tau(z=0),$$
$$p_\iota(z=0) - p_\rho(z=0) = p_\tau(z=0),$$
$$A_\iota \sin(\omega t) - A_\rho \sin(\omega t + \varphi_\rho) = A_\tau \sin(\omega t + \varphi_\tau). \tag{7.20}$$

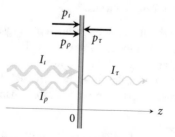

Figure 7.17: Part of incident sound wave (I_ι) is reflected (I_ρ) and part transmitted (I_τ).

Next we write Newton's second law for the panel. Panel acceleration a_{pan} is the result of the sound pressure difference between both sides of the panel as in

$$m\, a_{pan} = S\left[p_\iota(z=0) + p_\rho(z=0)\right] - S\, p_\tau(z=0),$$

where S is the panel area. Taking into account that $a_{pan} = dv_{pan}/dt$ and $\rho_A = m/S$, we get

$$p_\iota(z=0) + p_\rho(z=0) = \rho_A \frac{dv_{pan}}{dt} + p_\tau(z=0) = \frac{\rho_A}{\rho c}\frac{dp_\tau(z=0)}{dt} + p_\tau(z=0),$$

$$A_\iota \sin(\omega t) + A_\rho \sin(\omega t + \varphi_\rho) = \frac{\rho_A}{\rho c} A_\tau \omega \cos(\omega t + \varphi_\tau) + A_\tau \sin(\omega t + \varphi_\tau).$$

Inserting (7.20), we get

$$A_\iota \sin(\omega t) = \left[\frac{\rho_A \omega}{2\rho c} \cos(\omega t + \varphi_\tau) + \sin(\omega t + \varphi_\tau)\right] A_\tau,$$

$$A_\iota \sin(\omega t) = \sqrt{\left(\frac{\rho_A \omega}{2\rho c}\right)^2 + 1}\, A_\tau \sin(\omega t). \qquad (7.21)$$

We have used a trigonometric identity for the linear combination of sine and cosine, $a \sin x + b \cos x = \sqrt{a^2 + b^2}\sin(x + \varphi')$. Because $\varphi' = \arctan(b/a) = -\varphi_\tau$, the phase angle for the transmitted wave is

$$\varphi_\tau = -\arctan\left(\frac{\rho_A \omega}{2\rho c}\right).$$

The ratio of amplitudes is therefore

$$\frac{A_\tau}{A_\iota} = \frac{1}{\sqrt{\left(\frac{\rho_A \omega}{2\rho c}\right)^2 + 1}} = \frac{1}{\sqrt{\left(\frac{\pi \rho_A f}{\rho c}\right)^2 + 1}},$$

so the transmittance and sound reduction index (7.7) are

$$\tau(f) = \frac{I_\tau}{I_\iota} = \frac{|p_\tau^2|}{|p_\iota^2|} = \left(\frac{A_\tau}{A_\iota}\right)^2 = \frac{1}{\left(\frac{\pi \rho_A f}{\rho c}\right)^2 + 1},$$

$$R(f) = 10 \lg\left[\left(\frac{\pi \rho_A f}{\rho c}\right)^2 + 1\right]. \qquad (7.22)$$

For typical building elements $f \gg \rho c/\pi \rho_A$, so we get

$$R(f) \approx 20 \lg\left(\frac{\pi}{\rho c}\rho_A f\right) = 20 \lg(\rho_A f) - 42.4\,\text{dB}. \qquad (7.23)$$

The preceding expression is called the *mass law* of sound barriers. Observe that the sound reduction index increases with the surface density of the barrier: Doubling the surface density causes the increase of 6 dB. Therefore, the simplest method to increase the airborne sound reduction index of a building element or sound barrier is to increase its mass. For comparison, in standard DIN 4109-2 [9] weighted apparent sound reduction index (Section 7.4.2) for single massive wall is expressed as

$$R'_w = 28 \lg \rho_A - 18 \, \text{dB}. \tag{7.24}$$

It is instructive to observe two extreme solutions to equations (7.20) and (7.21). For a panel of negligible surface mass $R = 0$, $A_\rho = 0$, $A_l = A_\tau$ and $\varphi_\tau = 0$, which means that the complete sound is transmitted. For a panel of infinite surface mass, $R = \infty$, $A_\tau = 0$, $A_l = A_\rho$ and $\varphi_\rho = \pi$, which means that the complete sound is reflected.

For more strict consideration of sound transmission, including wave sound dissipation, wave functions are treated as complex functions of type $\exp[i(\omega t - kz)]$. However, those calculations are beyond the scope of this book.

7.4.2 Measuring and rating the airborne sound reduction index

The setup for measuring the airborne sound reduction index consists of two rooms—the source room and the receiving room—separated by the horizontal or vertical building element of interest. Both rooms must fulfil conditions for creating the diffuse sound field. The source room (left) contains the sound source and the first microphone, whereas the receiving room (right) contains the second microphone (Fig. 7.18). The sound source is positioned far from the building element and the first microphone, so the diffuse sound field conditions are met at their locations. The first and second microphone measure the sound pressure levels L_{p1} and L_{p2}, respectively.

We must also adapt expressions (7.3) and (7.8) for the diffuse sound field, in a similar way as we did in Section 7.2.2. To account for the fact that under diffuse field conditions the plane waves strike the building element uniformly from all directions, we have to add a one-forth factor:

$$I_\tau = \frac{1}{4} \tau I_1,$$

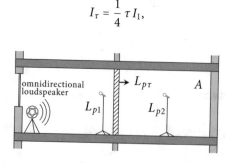

Figure 7.18: Measurement of the airborne sound reduction index for the hatched vertical building element. The source room (left) contains the sound source and the first microphone, whereas the receiving room (right) contains the second microphone.

where τ is transmittance of the room separating building element, whereas I and I_τ are the sound room and the transmitted sound intensities, respectively. By applying the logarithm we get

$$L_{p\tau} = L_{p1} - R + 10 \lg \frac{1}{4},$$

where R is sound reduction index of the building element and $L_{p\tau}$ transmitted sound pressure level. The sound reduction index is therefore

$$R = L_{p1} - L_{p\tau} - 10 \lg 4. \tag{7.25}$$

On the other hand, in the receiving room, the building element is actually a sound source, so diffuse sound field L_{p2} differs from the transmitted sound pressure level $L_{p\tau}$. Because the building element is the surface sound source with surface sound power level L_W'' (6.38), and, in its close proximity, the sound pressure level is $L_{p\tau}$, we can use expressions (6.40) and (6.41) to establish

$$L_W'' = L_{p\tau},$$
$$\implies L_W = L_{p\tau} + 10 \lg \frac{S}{S_0},$$

where L_W is total sound power and S the area of the room separating building element.

Inserting this into expression (7.15), we finally obtain

$$L_{p2} = L_{p\tau} + 10 \lg \frac{S}{S_0} - 10 \lg \frac{A}{4S_0} = L_{p\tau} + 10 \lg \left(\frac{4S}{A} \right), \tag{7.26}$$

where A is the equivalent absorption area of the receiving room.

Combining (7.25) and (7.26), for the sound reduction index, we get

$$R = L_{p1} - L_{p2} + 10 \lg \left(\frac{S}{A} \right) \quad \text{(laboratory measurement)}. \tag{7.27}$$

Unlike expression (7.8), this expression is valid in diffuse sound field conditions, taking into account reflections on both sides of the sound barrier.

The exact airborne sound reduction index can be obtained only by laboratory measurements, prescribed by standard ISO 10140-2 [36]. Note that we strive for an airborne sound reduction index that is *as large as possible*.

The airborne sound reduction index also can be obtained by the *field measurement*, that is, the *in situ* measurement of the building element in the constructed dwelling, prescribed by standard ISO 16283-1 [48]. However, the result of the measurement is influenced not only by airborne sound transmission but also by flanking sound transmissions and acoustic bridges. In this case, we obtain the *apparent sound reduction index* as

$$R' = L_1 - L_2 + 10 \lg \left(\frac{S}{A} \right) \quad \text{(field measurement)}, \tag{7.28}$$

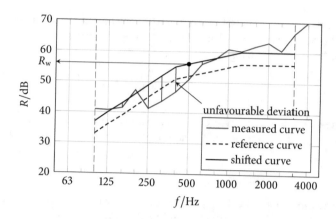

Figure 7.19: Procedure for determining the airborne sound reduction index rating R_w specified by standard ISO 717-1 [23]. The reference curve is shifted towards measured curve, after which the value of the shifted curve at 500 Hz is read out.

which is obviously smaller than the one obtained by laboratory measurements $R' < R$.

Finally, the preceding expression can be further simplified using the Sabine equation (7.18) to obtain the *standardised level difference* as

$$D_{nT} = L_1 - L_2 + 10 \lg \left(\frac{T_{60}}{T_0} \right), \tag{7.29}$$

where T_0 (s) is the *reference reverberation time.* For dwellings, $T_0 = 0.5\,\text{s}$, which corresponds to characteristic room dimension $V/S = T_0/(0.163\,\text{s/m}) = 3.1\,\text{m}$.

We know from the mass law (7.23) that the obtained (apparent) sound reduction index and standardised level difference are strongly frequency dependent and increase for about 6 dB per octave (doubling of frequency). However, standard ISO 717-1 [23] enables the airborne sound reduction index to be rated with a *single value.* The standard defines a reference curve (Fig. 7.19), which is shifted towards a measured curve in 1 dB steps until the total of the unfavourable deviations (negative difference between the measured curve and the shifted curve) is as close as possible to, but not greater than 32 dB. The value of the shifted curve at 500 Hz is *weighted sound reduction index* R_w, *weighted apparent sound reduction index* R'_w or *weighted standardised level difference* $D_{nT,w}$, depending on the input data.

7.4.3 Measuring and rating the impact sound pressure level

Setup for measuring the impact sound pressure level consists of the source room and receiving room, separated by the horizontal building element of interest. The receiving room must fulfil the conditions for creating the

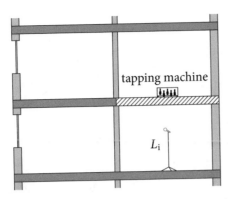

Figure 7.20: Measurement of the impact sound pressure level for the hatched building element. The tapping machine is on the floor of the source room (top), whereas the receiving room (bottom) contains the microphone.

diffuse sound field. The standardised tapping machine, having five 500 g hammers that free-fall from a height of 40 mm in 100 ms intervals [38], is put on the floor of the source room [49]. The microphone is positioned in the receiving room and measures *impact sound pressure level* L_i (dB) (Fig. 7.20).

The impact sound pressure level depends not only on properties of the building element but also on the equivalent absorption area of receiving room A (7.15). To remove the influence of the latter, we define *normalised impact sound pressure level* L_n as

$$L_n = L_i + 10 \lg \left(\frac{A}{A_0} \right) \quad \text{(laboratory measurement),} \qquad (7.30)$$

where A_0 (m^2) is the reference absorption area. For dwellings, $A_0 = 10 \, \text{m}^2$.

The exact normalised impact sound pressure level can be obtained only by laboratory measurements, prescribed by standard ISO 10140-3 [37]. Note that we strive for an normalised impact sound pressure level that is *as small as possible*, in contrast to the airborne sound reduction index.

The normalised impact sound pressure level also can be obtained by the *field measurement*, that is, the *in situ* measurement of the building element in the constructed dwelling, prescribed by standard ISO 16283-2 [49]. However, the result of the measurement is influenced not only by impact sound transmission but also by flanking sound transmissions and acoustic bridges. For field measurements, an apostrophe is added to the symbol:

$$L'_n = L_i + 10 \lg \left(\frac{A}{A_0} \right) \quad \text{(field measurement),} \qquad (7.31)$$

The normalised impact sound pressure level obtained by field measurements is obviously larger than the one obtained by laboratory measurements $L'_n > L_n$.

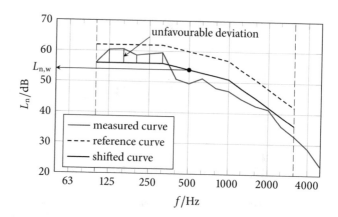

Figure 7.21: Procedure for determining the impact sound pressure level rating $L_{n,w}$ specified by standard ISO 717-2 [24]. The reference curve is shifted towards measured curve, after which the value of the shifted curve at 500 Hz is read out.

Finally, the preceding expression can be further simplified using the Sabine equation (7.18) to obtain *standardised impact sound pressure level* L_{nT} as

$$L_{nT} = L_i - 10 \lg\left(\frac{T_{60}}{T_0}\right), \qquad (7.32)$$

where T_0 (s) is the *reference reverberation time.* For dwellings, $T_0 = 0.5\,\mathrm{s}$, which corresponds to characteristic room volume $V = (A_0\,T_0)/(0.163\,\mathrm{s/m}) = 31\,\mathrm{m}^3$.

Just as in the case of the airborne sound reduction index, the obtained normalised and standardised impact sound pressure levels are strongly frequency dependent. However, standard ISO 717-2 [24] enables impact sound pressure level to be rated with a *single value*. The standard defines a reference curve (Fig. 7.21), which is moved towards the measured curve in 1 dB steps until the total of the favourable deviations (negative difference between the shifted curve and the measured curve) is as close as possible to, but not greater than 32 dB. The value of the shifted curve at 500 Hz is either *weighted normalised impact sound pressure level* $L_{n,w}$, $L'_{n,w}$ or *weighted standardised impact sound pressure level* $L_{nT,w}$, depending on the input data.

> **Info box**
>
> Standard ISO 717-2 is used to rate impact sound pressure level with a single value.

Minimum values for weighted apparent sound reduction index R'_w and maximum values for weighted normalised impact sound pressure level (field measurement) $L'_{n,w}$ are the matter of state legislation. For orientation, some limit values for internal building elements from standard DIN 4109-1 [8] are specified in Table 7.1. Minimum values of weighted apparent sound reduction index R'_w for external building elements are also specified depending on the sound pressure level at the façade and the type of room. Standard DIN 4109-2 [9] and its dependencies also provide precise calculation methods for weighted apparent sound reduction index and weighted normalised impact sound pressure level (field measurement), which is however beyond the scope of this book.

Table 7.1: Limit values for weighted apparent sound reduction index R'_w and weighted normalised impact sound pressure level (field measurement) $L'_{n,w}$ from standard DIN 4109-1 [8].

building element	R'_w/dB	$L'_{n,w}/dB$
walls to rooms with noisy building services	57, 62	—
floors to rooms with noisy building services	—	43
floors to other dwellings	54	50
walls to other dwellings	53	—
doors from corridor of dwellings to stairways	27	—
doors from living room of dwellings to stairways	37	—

Problems

7.1 A receiver and transmitter of height 1.5 m are located in a hall of rectangular floor plan of dimensions 30 m × 40 m and height 6 m, as shown in the figure. Calculate the time delays for sound waves reflected from the right wall and sound waves reflected from the ceiling. (53 ms, 3.6 ms)

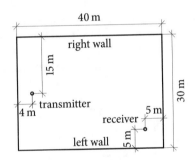

7.2 The sound pressure level of a sound that is reflected once should be smaller for 7.0 dB. Calculate the minimal absorbance of the wall. (0.8)

7.3 Sound is reflected twice from the walls of absorbance 0.2. What is the sound pressure level reduction? Neglect the reduction due to the sound path. (1.9 dB)

7.4 Calculate the sound pressure level difference between the direct and both reflected sound waves in problem 7.1. Do the reflected sound waves together have a larger sound pressure level than the direct sound wave? For how much? Take into account the geometric divergence, and take the absorbance of the walls and ceiling to be 0.10. (4.28 dB, 0.78 dB, yes, 0.82 dB)

7.5 A lecture room of rectangular floor plan with dimensions 10 m × 20 m and height 4 m, has concrete walls and ceiling, parquet on the floors, six windows each of area 10 m² and 70 seats. Calculate the reverberation time. Take the absorbance of concrete to be 0.02, absorbance of timber to be 0.06, absorbance of glass to be 0.03 and equivalent absorption area of each empty seat to be 0.5 m². (2.3 s)

7.6 A person in the open space is listening to the point sound source located 40 m away. At some moment, the barrier with a sound reduction index of 5.0 dB is positioned between the person and the sound source, perpendicularly to the sound path. At what distance from the sound source should the person move in order to hear the sound as loud as in the beginning? (22.5 m)

7.7 The sound source of sound power level 80 dB is located in the sound room. What are the diffuse sound pressure levels in the sound room and the neighbouring room? The area of the wall between the rooms is 8 m², and the sound reduction index is 50 dB. Take the equivalent absorption area of the sound rooms to be 12 m² and the equivalent absorption area of the neighbouring rooms to be 20 m². (75 dB, 21 dB)

7.8 Two rooms of floor plan dimensions $l_1 \times w_1$ = 5.0 m × 3.0 m and $l_2 \times w_2$ = 3.0 m × 2.5 m and of a height 2.5 m are positioned as shown in the figure. Reverberation times measured in both rooms are T_1 = 0.40 s and T_2 = 0.25 s. After we turn on the sound source in the first room, we find that diffusive sound pressure levels are L_1 = 82 dB and L_2 = 39 dB. What is the sound reduction index of the wall between the rooms? (40.9 dB)

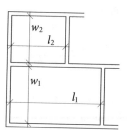

7.9 The horizontal distance between the right traffic lane and the building is 5.3 m and between the right traffic lane and the transparent noise barrier is 2.9 m, as shown in the figure. Calculate the sound pressure level on the façade of the building 4.0 m above the ground, if absorbance of the barrier is 0.20. Assume for the traffic lane that the sound source is 5.0 cm above the ground and take the linear sound power level to be 70 dB. Take into account the direct sound and the sound reflected by the barrier. (55.4 dB)

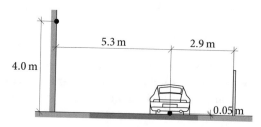

7.10   A living room of rectangular foor plan of dimensions $l_2 \times w_2 =$ 5.0 m $\times$ 3.0 m is located directly adjacent to the engine room of rectangular foor plan of dimensions $l_1 \times w_1$ = 2.0 m $\times$ 2.0 m, as shown in the figure. Both rooms are of height 2.6 m. The sound reduction index between the rooms is 57 dB, whereas the reverberation time of the living room is 0.50 s. What is the highest acceptable sound pressure level in the engine room, if the highest permissible sound pressure level in the room is 35 dB? There is a device of sound power 100 dB in the engine room. In order not to exceed the highest acceptable sound pressure level, sound absorbers are installed on all room surfaces. What is their smallest absorbance? Neglect the increase of the sound reduction index due to the installation of absorbers. (95.9 dB; 0.36)

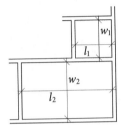

8 Illumination

Illumination is the building physics subject with a crucial influence on human life. Indoor illumination can improve of work efficiency, comfort, orientation and health. Illumination from natural sources or daylighting stands out as the most economic means that also enhances our sense of time and increases the economic value of the building. On the other hand, outdoor illumination provides traffic and population safety. As we will see in this chapter, light phenomena are similar to those mentioned in previous chapters, although there are also some important differences.

8.1 Introduction

In Chapters 2 and 3 we dealt with electromagnetic waves (radiation), but only as a heat transfer mechanism. In this section, we start with *radiometry*, which studies propagation and measurement of electromagnetic radiation more precisely. Later, we will look more closely at light, which is the most important part of the spectrum that extends from approximately 400 nm to approximately 750 nm. In contrast to the previous approach, we will be concerned with radiation and light as *wave phenomena*, and we will study human eye sensitivity as a function of wavelength.

First we will scarcely reproduce a derivation of electromagnetic wave functions, showing a few of the most basic steps without mathematical formalism. A strict and exact procedure is beyond the scope of this book and should be located elsewhere. We start with two Maxwell equations in the vacuum

$$\vec{\nabla} \times \vec{E} = -\frac{\partial \vec{B}}{\partial t},$$

$$\vec{\nabla} \times \vec{B} = \mu_0 \varepsilon_0 \frac{\partial \vec{E}}{\partial t},$$

where E (V/m) is the electric field, B (T) is the magnetic field, $\mu_0 = 1.256 \times 10^{-6}$ H/m permeability of vacuum and $\varepsilon_0 = 8.854 \times 10^{-12}$ F/m permittivity of vacuum.

For an electromagnetic wave travelling in the z-direction with an electrical field in the x-direction E_x and a magnetic field in the y-direction B_y, the

M. Pinterić, *Building Physics*,
https://doi.org/10.1007/978-3-030-67372-7_8

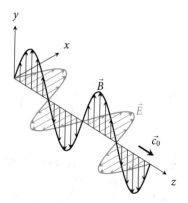

Figure 8.1: Linearly polarised electromagnetic wave. The electric field changes in x-direction and the magnetic field changes in y-direction as the wave propagates in z-direction.

equations for electric and magnetic fields can be decoupled into the following equations:

$$\frac{\partial^2 E_x}{\partial t^2} = \frac{1}{\mu_0 \varepsilon_0} \frac{\partial^2 E_x}{\partial z^2},$$

$$\frac{\partial^2 B_y}{\partial t^2} = \frac{1}{\mu_0 \varepsilon_0} \frac{\partial^2 B_y}{\partial z^2}.$$

Comparing the expressions with the wave equation (5.16), we conclude that one possible solution is a linearly polarised electromagnetic wave (5.7) as shown in Fig. 8.1 and as

$$E_x(z, t) = E_0 \sin(\omega t - kz),$$

$$B_y(z, t) = B_0 \sin(\omega t - kz),$$

where E_0 and B_0 are electric and magnetic field amplitudes, respectively, and the *speed of light* is

$$c_0 = \frac{1}{\sqrt{\mu_0 \varepsilon_0}}. \tag{8.1}$$

The capability of electromagnetic waves to transfer energy is described by the density of energy flow rate called *Poynting vector* S (W/m^2) and defined as

$$\vec{S} = \frac{1}{\mu_0} \vec{E} \times \vec{B}. \tag{8.2}$$

Electric and magnetic fields are changing rapidly—oscillating with rather large frequency—so the temporary density of the energy flow rate is neither easily observable nor a very useful quantity. We are therefore more interested in the time-averaged value called *irradiance* E (W/m^2):

$$E = \langle S \rangle = \frac{1}{t'} \int_0^{t'} S \, dt. \tag{8.3}$$

Irradiance E and *radiative* density of heat flow rate q (2.3) are actually the same quantity.

Another important quantity is the emitted energy flow rate called *radiant flux* Φ (W). Radiant flux is actually the same quantity as *radiative* heat flow rate Φ (2.2).

The study of electromagnetic waves energy resembles closely the study of sound wave energy (Section 6.1.2). Poynting vector S, irradiance E and radiant flux Φ directly correspond to sound intensity i (6.9), time-averaged sound intensity I (6.12) and sound power P (6.22), respectively.

> **Info box**
>
> Radiant flux and radiance are the same quantities as radiative heat flow rate and radiative density of heat flow rate, respectively.

8.2 Optical properties of materials

In this section, we will recapitulate some facts already presented in Section 2.4.1 on page 56, but we will consider a few additional details that are of interest when studying light propagation.

As we pointed out previously, when radiation (light) strikes liquid or solid material, part of the incident radiation is absorbed, part is transmitted and part is reflected (Fig. 8.2). As discussed in Section 2.4 on page 53, the radiant (light) spectrum is specified by wavelength. If we denote radiant flux for incident radiation Φ, for reflected radiation Φ_ρ, for absorbed radiation Φ_α and for translated radiation Φ_τ, we can define *reflectance* ρ, *absorptance* α and *transmittance* τ as

$$\rho(\lambda) = \frac{\Phi_\rho(\lambda)}{\Phi(\lambda)}, \tag{8.4}$$

$$\alpha(\lambda) = \frac{\Phi_\alpha(\lambda)}{\Phi(\lambda)}, \tag{8.5}$$

$$\tau(\lambda) = \frac{\Phi_\tau(\lambda)}{\Phi(\lambda)}. \tag{8.6}$$

Here we took into account that all three physical quantities depend on the radiation (light) wavelength. Note that their value range is $0 \le \rho(\lambda), \alpha(\lambda), \tau(\lambda) \le 1$.

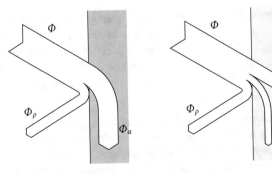

Figure 8.2: Radiant (light) processes for nontransparent (left) and transparent (right) objects. Part of the incident radiant flux (Φ) is reflected (Φ_ρ), part is absorbed (Φ_α) and part is transmitted (Φ_τ).

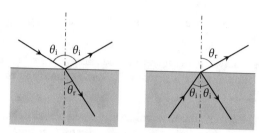

Figure 8.3: Snell's law. Angles of incidence and reflection θ_i are equal, whereas the angle of refraction θ_r depends on the angle of incidence and speeds of light in both substances. The left side depicts the light beam incident on a substance with lower speed of light (usually denser substance), and the right side light beam incident on a substance with a higher speed of light (usually thinner substance).

Because energy is conserved in the process, the sum of radiant fluxes for reflected, absorbed and translated radiation should be equal to the radiant flux for incident radiation:

$$\Phi_\rho(\lambda) + \Phi_\alpha(\lambda) + \Phi_\tau(\lambda) = \Phi(\lambda).$$

Together with the definition of reflectance, absorptance and transmittance, this leads to

$$\rho(\lambda) + \alpha(\lambda) + \tau(\lambda) = 1. \tag{8.7}$$

In light propagation, we are not only interested in energy relations but also in how light is propagating after striking a substance. We start this study by defining the *normal line* as the line perpendicular to the surface at the point that the light ray strikes or leaves the substance (usually drawn as a dash-dotted line). We define angles of incidence, reflection and refraction as angles between the normal line and incident, reflected and refracted rays, respectively. As it turns out, angles of incidence and reflection are equal, whereas angles of incidence and refraction are connected by Snell's law (Fig. 8.3) as in

$$\frac{\sin \theta_1}{\sin \theta_2} = \frac{c_1}{c_2}, \tag{8.8}$$

where c_1 and c_2 are speeds of light in corresponding media. Snell's law is the basic law of optics.

An interesting phenomenon occurs for light beams striking a substance with higher speed of light (usually thinner substance) at large incident angles (Fig. 8.3, right). If the calculated angle of reflection is $\theta_r \geq 90°$, all of the light is reflected. This phenomenon is called *total reflection* and is used to transmit light within optical fibres with negligible loss or to create light transparent shading devices.

In many cases, however, we are not interested in rays that are refracted into the object, but in rays that are transmitted through the object. For thin transparent objects with plane parallel surfaces (windows), angles of incidence, reflection and transmission outside the object for a single ray are always equal. However, the reflection and transmission of the whole beam also depend on the structure of the surface:

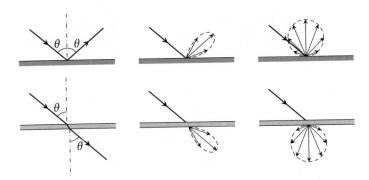

Figure 8.4: Specular reflection and transmission (left), glossy reflection and transmission (middle) and diffuse reflection and transmission (right).

- One class of materials has almost perfectly *smooth surfaces*, even at the microscopic level, for example, mirrors, clear glass and calm bodies of water. As such, they offer each individual ray of the beam the same surface orientation, and all reflected rays have the same direction, forming a reflected beam. Reflections and transmissions on smooth surfaces are known as *specular* reflections and transmissions (Fig. 8.4, left).

- Another class of materials has *rough surfaces*, for example, walls or ground glass. Each individual ray of the beam strikes at a part of the surface that is oriented differently. Consequently, reflected rays have various directions, and no beam is formed. Reflections and transmissions on rough surfaces are known as *diffuse* reflections and transmissions (Fig. 8.4, right).

- Many materials have properties between those two extremes. Each individual ray of the beam strikes at a part of the surface that is oriented differently, but the reflected rays are still headed closely in the direction of the ideal specular reflection. Reflections and transmissions on such intermediate surfaces are known as *glossy* reflections and transmissions (Fig. 8.4, middle).

8.3 Photometry

Photometry is the discipline that studies light in terms of its perceived brightness to the human eye. In this section, we will introduce photometric quantities and explain how they describe characteristics of human visual perception.

8.3.1 Wavelength perception

We have already mentioned that visual light corresponds to radiation between approximately 400 nm and approximately 750 nm. However, even within that range, not all wavelengths are perceived as equally bright.

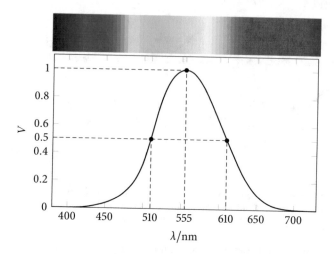

Figure 8.5: The spectral luminous efficiency $V(\lambda)$ [72] describes the wavelength-dependent sensitivity of the human eye. The eye is most sensitive at 555 nm, for which $V = 1$.

The human eye is actually most sensitive for wavelengths of 555 nm. Of all light wavelengths, this one gives the impression of greatest brightness for the same radiant flux.

In order to compensate for the wavelength-dependent sensitivity of human vision, separate units and quantities were defined historically. In place of radiant flux Φ for the complete radiant spectrum, we define *luminous flux* with unit *lumen* Φ_v (lm) for the visual light spectrum. In particular, at wavelength 555 nm, radiant flux $\Phi = 1\,\text{W}$ produces luminous flux $\Phi_v = 683\,\text{lm}$. We write

$$\Phi_v(555\,\text{nm}) = K_m\,\Phi(555\,\text{nm}), \tag{8.9}$$

where $K_m = 683\,\text{lm/W}$ is the maximum spectral luminous efficacy. Quantity luminous flux with unit lumen (lm) therefore describes the amount of visible light emitted by a source.

Sensitivity at other wavelengths is described by spectral luminous efficiency $V(\lambda)$ (Fig. 8.5). The function has maximum value $V = 1$ at 555 nm, which is the wavelength of the highest eye sensitivity. Using this function, the relation between radiant flux and luminous flux for a *single wavelength* can be generalised as

$$\Phi_v(\lambda) = K_m\,V(\lambda)\,\Phi(\lambda). \tag{8.10}$$

For example, at 510 nm or 610 nm, the value is $V = 0.5$, which implies that radiant flux $\Phi = 1\,\text{W}$ at these wavelengths 'produces' luminous flux $\Phi_v = 342\,\text{lm}$. On the other hand, radiant flux $\Phi = 2\,\text{W}$ at 510 nm or 610 nm 'produces' the same brightness as radiant flux $\Phi = 1\,\text{W}$ at 555 nm.

Common light sources emit a wide light spectrum, so we have to integrate over all wavelengths as

$$\Phi_v = K_m \int \frac{\mathrm{d}\Phi}{\mathrm{d}\lambda}\,V(\lambda)\mathrm{d}\lambda. \tag{8.11}$$

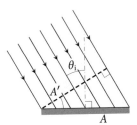

Figure 8.6: Illuminance for a surface of area A and angle of incidence θ_i. When light does not strike the surface perpendicularly, an element of the surface intercepts smaller luminous flux and appears darker.

where $d\Phi/d\lambda$ is the spectral radiant flux. Luminous flux is always provided for electric bulb products as the most important specification (Fig. 8.18 on page 265).

In place of irradiance E for the complete radiant spectrum, we define *illuminance* with unit *lux* E_v (lx) for visual light:

$$E_v = K_m \int \frac{dE}{d\lambda} V(\lambda) d\lambda. \tag{8.12}$$

Illuminance describes human perception of brightness in a similar way that the A-weighted sound pressure level L_{pA} describes human perception of loudness.

> **Info box**
>
> Luminous flux and illuminance are defined to compensate for wavelength-dependent sensitivity of human vision.

For energy transfer from an isotropic point light source, the illuminance is, similar to (6.30), expressed as

$$E_v = \frac{\Phi_v}{A} = \frac{\Phi_v}{4\pi r^2}. \tag{8.13}$$

Obviously, unit lux and lumen are connected, $lx = lm/m^2$.

Illuminance is often defined (hence its name) to describe the brightness of surfaces at room temperature. The surfaces do not radiate light themselves, so they appear bright as a result of reflected light. Brightness will be proportional to the reflected luminous flux *per surface area*, which is proportional to the incident luminous flux *per surface area* (8.13). When incident light rays are not perpendicular to the surface (Fig. 8.6), the effective luminous flux surface A' is reduced to

$$A' = A \cos \theta_i,$$

where θ_i is the angle of incidence. For surface, the illumination expression is then transformed to

$$E_v = \frac{\Phi_v}{A} = \frac{\Phi_v}{A'} \cos \theta_i = \frac{\Phi_v}{4\pi r^2} \cos \theta_i. \tag{8.14}$$

8.3.2 Directivity perception

The most important difference between auditory and visual sensations refers to the fact that the visual sensation also has *spatial resolution*. One

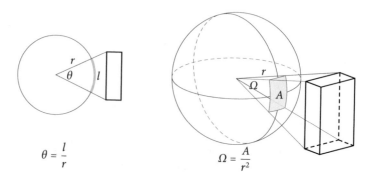

$$\theta = \frac{l}{r} \qquad\qquad \Omega = \frac{A}{r^2}$$

Figure 8.7: The angle θ value is by definition the ratio of the circle arc length l representing part of the two-dimensional space to the circle radius r (left). We can measure the angle of a two-dimensional object by projecting it on an imaginary circle. Similarly, the solid angle Ω value is by definition the ratio of the sphere segment area A representing part of the three-dimensional space to the square of the sphere radius r (right). We can measure the solid angle (apparent visual size) of an three-dimensional object by projecting it on an imaginary sphere.

ear only detects the frequency and density of the sound energy flow rate (intensity), whereas one eye detects wavelength, density of light energy flow rate, as well as *direction* and *visual size* of the light source. Incidentally, two ears can also detect the direction of the sound source using the time delay of the sound wave between two ears, whereas two eyes can also tell the distance of the light source from the difference in their orientation.

Furthermore, ordinary light sources are not isotropic, and luminous flux depends on *direction*. Additional 'complications' can be described in terms of *solid angle* Ω.

We are already well acquainted with the concept of the angle, which represents a part of two-dimensional space delimited by two rays sharing a common endpoint. The angle value is defined as the ratio of the circle arc length representing the two-dimensional space to the circle radius r (Fig. 8.7, left) as in

$$\theta = \frac{l}{r}. \tag{8.15}$$

The angle is dimensionless, however standard unit is called *radian* (rad). Other units, such as degrees, are also widely used. Note that the angle of the full two-dimensional space is equal to 2π.

On the other hand, the solid angle represents a part of three-dimensional space delimited by several rays sharing a common endpoint. The solid angle value is by definition the ratio of the sphere segment area representing part of the three-dimensional space to the square of the sphere radius r (Fig. 8.7, right) as in

$$\Omega = \frac{A}{r^2}. \tag{8.16}$$

For example, the apparent visual size of an arbitrary object (proportion of total sight 'occupied' by the object) can be directly quantified by the

Figure 8.8: Luminous intensity of two incandescent bulbs of the same electrical power and luminous flux, classic isotropic light bulb (left) and reflector light bulb (right). With the reflector light bulb, the light is concentrated into the smaller solid angle, which increases the luminous intensity there. Note that the luminous intensity of the reflector bulb also depends on the deflection angle.

solid angle, which is calculated by projecting the object onto imaginary sphere.

The solid angle is dimensionless, however, the unit *steradian* (sr) is often used to distinguish between dimensionless quantities with different natures. The solid angle of the full three-dimensional space is equal to 4π.

We have already came across the solid angle when we considered the directivity of nonisotropic sound sources (see Section 6.2.4 on page 199). The same approach is used for nonisotropic light sources, which are light sources that emit light only in certain directions or for which light emission is direction dependent (Fig. 8.8). We define *luminous intensity* with unit *candela* I_v (cd) as the quotient of luminous flux by the solid angle into which light is radiated:

$$I_v = \frac{\Phi_v}{\Omega}.$$ (8.17)

Note that the unit of luminous intensity, candela cd = lm/sr, is one of seven basic SI units. As shown in Fig. 8.8 (right), when luminous flux is directed into a smaller solid angle, the luminous intensity increases.

The definition of luminous intensity leads to a generalisation of (8.13) and (8.14) for nonisotropic light sources by replacing full solid angle 4π by Ω. For the energy transfer, we get

$$E_v = \frac{I_v}{r^2},$$ (8.18)

whereas for the surface, we get

$$E_v = \frac{I_v}{r^2} \cos\theta_i.$$ (8.19)

Figure 8.9: The illumination from a more distant light source is smaller, yet the light source appears *equally bright* and *smaller*.

Next we discuss the *spatial resolution* of visual sensation. When an observer moves away from the *sound* source, the sound appears *less loud*. We can conclude that for our brains, the loudness corresponds to the sound intensity, which also decreases with distance. On the other hand, when an observer moves away from the *light* source, the source appears *smaller* but *equally bright* (Fig. 8.9). This means that for our brains, brightness does not correspond to illumination, which decreases with distance. To account for this property of visual sensation, we define a new physical quantity called *luminance* L_v (cd/m^2) as the quotient of illumination by the solid angle as in

Info box
Luminance is based on the fact that light sources appear equally bright regardless of the distance.

$$L_v = \frac{E_v}{\Omega}.$$ (8.20)

We will now use a simple example to show that luminance for an isotropic light source is essentially independent of the distance. As shown in Fig. 8.10, when the distance between the observer and source increases, both the solid angle as

$$\Omega \approx \frac{\pi R^2}{r^2}$$

and the illumination (8.18) decrease, so their quotient, luminance, is constant:

$$L_v \approx \frac{I_v}{r^2} \frac{r^2}{\pi R^2} = \frac{I_v}{\pi R^2}.$$ (8.21)

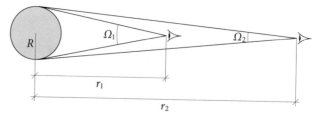

Figure 8.10: Luminance of the spherical isotropic light source is essentially independent of the distance between the source and the observer. Both illumination and solid angle decrease with distance, so their quotient is constant.

For the arbitrary shape of the light source, we get

$$L_v = \frac{E_v}{\Omega} = \frac{I_v \, r^2}{r^2 \, A},$$

$$L_v = \frac{I_v}{A}, \tag{8.22}$$

where A is the sphere segment of the light source. Therefore, luminance is the quotient of luminous intensity I_v by area of orthogonal projection of the light source on a plane perpendicular to the light direction A. This is the official definition of luminance and is obviously independent of the light receiver position. Luminance is always provided for computer and television screens as the most important specification.

Note that both luminous intensity and luminance are expressed in terms of candela. Candela-based units imply that the quantity is normalised in regard to the solid angle.

Example 8.1: Sun's photosphere.

In Example 2.1, we have calculated the heat flow rate, or radiant flux, of the Sun. If the spectral luminous efficacy of the Sun is $K = 93 \, \mathrm{lm/W}$, calculate its luminous flux and luminous intensity. Calculate the luminance of the Sun's *photosphere—visual surface—* at the outer edge of the Earth's atmosphere, if the radius of the Sun is $r_{Sun} = 6.96 \times 10^5 \, \mathrm{km}$.

Luminous flux and intensity are (8.17)

$$\Phi_v = K \, \Phi_{Sun} = 3.58 \times 10^{28} \, \mathrm{lm},$$

$$I_v = \frac{\Phi_v}{\Omega} = \frac{\Phi_v}{4\pi} = 2.85 \times 10^{27} \, \mathrm{cd},$$

where we assumed that the Sun is the isotropic light source. In order to calculate luminance (8.22), we have to find the orthogonal projection of the Sun on a plane perpendicular to the light direction. This is simply the area of the circle with the radius of the Sun:

$$L_v = \frac{I_v}{A} = \frac{I_v}{\pi r_{Sun}^2} = 1.87 \times 10^9 \, \frac{\mathrm{cd}}{\mathrm{m}^2}.$$

Note that luminance is independent of the distance from the light source; therefore, we do not need the distance between the Sun and the Earth. Furthermore, note that luminance is somewhat smaller at the surface of the Earth because the atmosphere absorbs and reflects part of the incident light. At zenith a on sunny day, luminance observed at the Earth's surface is about $1.44 \times 10^9 \, \mathrm{cd/m}^2$.

Illumination also can be calculated from luminance using expression (8.20). For energy transfer, we get

$$E_v = L_v \, \Omega, \tag{8.23}$$

whereas for the surface, we have to take into account the nonzero angle of incidence θ_i, as we have done in Section 8.3.1

$$E_v = L_v \, \Omega \cos \theta_i. \tag{8.24}$$

8.3.3 Diffusive light sources

Whereas in the previous sections we were primarily concerned with point light sources, now we turn our attention to transparent and reflective surfaces. Their treatment depends on their properties (Fig. 8.4 on page 247):

- For smooth surfaces and specular reflections and transmissions (mirror, clear glass), we approach the problem in terms of geometric optics. For reflections, the calculation is made in terms of the virtual image, that is, the source reflected over the mirror plane (Fig. 8.11, top left). For light transmissions, on the other hand, the effect of thin flat glasses can be neglected (Fig. 8.11, top right).

- For rough surfaces and diffusive reflections and transmissions (wall, ground glass), the observer perceives the object as a surface or diffusive light source (Fig. 8.11, bottom). In this section, we will take a closer look at their properties.

Note that reflectance of the rough surfaces is usually called *albedo* ('whiteness' in Latin). Albedo takes a wide range of values from 0.9 for fresh snow to 0.1 for asphalt.

One important property of most surface sources is that they appear equally bright regardless of the direction of the observation, defined by angle of

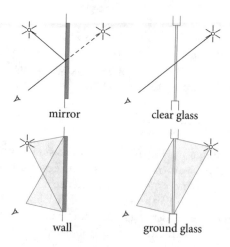

Figure 8.11: Optical treatment of surfaces. For smooth surfaces (top) the problem is treated in terms of geometric optics. For rough surfaces (bottom), objects are perceived as surface or diffusive light sources.

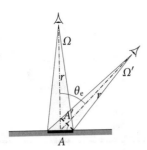

Figure 8.12: Brightness and therefore luminance are independent of angle of emittance θ_e. The fact that Ω' is smaller than Ω leads to luminous intensity angle dependence described by Lambert's law.

emittance θ_e (Fig. 8.12). When the angle of emittance increases, the solid angle decreases as in

$$\Omega = \frac{A'}{r^2}.$$

However, from (8.20)

$$L_v = \frac{E_v}{\Omega} = \text{constant},$$

we see that illuminance also must decrease in order for luminance to be constant. Using the expression for nonisotropic light sources (8.18), we get

$$L_v \approx \frac{I_v}{r^2}\frac{r^2}{A'} = \frac{I_v}{A\cos\theta_e} = \text{constant}.$$

Therefore, luminous intensity can be written as

$$I_v(\theta_e) = I_v(0)\cos\theta_e. \tag{8.25}$$

This statement is called *Lambert's law*.

> **Info box**
>
> Lambertian light sources appear equally bright regardless of the direction of the observation.

In order to calculate the total luminous flux of a small surface of area A, we divide the half-space into small differential solid angles of constant θ_e, as shown in Fig. 8.13. Because the radius is equal to $r = 1$, we get

$$d\Omega = 2\pi x\, ds = 2\pi \sin\theta_e\, d\theta_e.$$

The total luminous flux is then obtained from (8.17) and (8.25) as

$$\Phi_v = \int I_v\, d\Omega = \int_0^{\pi/2} I_v(0)\cos\theta_e\, 2\pi \sin\theta_e\, d\theta_e = \pi I_v(0). \tag{8.26}$$

Because the luminous flux is proportional to the surface area, it is common to define *luminous exitance* M_v [lm/m²] as the quotient of luminous flux by surface area as in

$$M_v = \frac{\Phi_v}{A}, \tag{8.27}$$

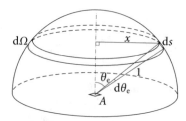

Figure 8.13: A differential ring of a half-sphere with constant angle of emittance above a small surface.

for which the differential form is

$$M_v = \frac{d\Phi_v}{dA}.$$ (8.28)

This transforms Lambert law into the following form

$$I_v(\theta_e) = \frac{1}{\pi} A M_v \cos\theta_e.$$ (8.29)

An overview of the illumination calculation procedures is shown below:

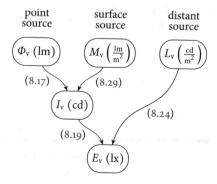

8.3.4 View factor

In Section 8.3.3 we saw that perfectly diffuse surfaces reflect light according to Lambert's law (8.25). This law also applies to for ideal diffuse radiant and light sources and must be taken into account when calculating the view factor, as defined in Section 2.4.2 on page 59. We will now derive the strict mathematical expression for the view factor calculation.

Let's observe the luminous flux from small surface A_i to small surface A_j (Fig. 8.14). Whereas the total luminous flux from small surface A_i is equal to $\pi I_{v,i}(0)$ (8.26), the luminous flux towards surface A_j that encompasses small solid angle Ω_{ij} for the angle of emittance θ_i is equal to (8.17)

$$\Phi_{v,ij} = \Omega_{ij} I_{v,i}(\theta_i) = \Omega_{ij} I_{v,i}(0) \cos\theta_i.$$ (8.30)

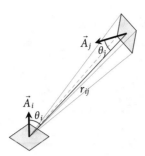

Figure 8.14: Luminous flux from small surface A_i to small surface A_j. Luminous flux depends on the solid angle Ω_{ij} that surface A_j encompasses and angle of emittance θ_i.

Because view factor F_{ij} is the ratio of light radiated towards surface A_j to total radiated light we get (8.26)

$$F_{ij} = \frac{\cos \theta_i \, \Omega_{ij}}{\pi}. \tag{8.31}$$

Using the definition of solid angle (8.16) and noting that we have to take into account only the projection of surface A_j on the direction towards surface A_i, that is, $A_j \cos \theta_j$, we finally get

$$F_{ij} = \frac{\cos \theta_i \, \cos \theta_j}{\pi \, r_{ij}^2} A_j. \tag{8.32}$$

For large surfaces, each surface must be decomposed into infinitesimal pieces, and the view factor is obtained by integration over both surfaces:

$$F_{ij} = \frac{1}{A_i} \int_{A_i} \int_{A_j} \frac{\cos \theta_i \, \cos \theta_j}{\pi \, r_{ij}^2} \, dA_i \, dA_j. \tag{8.33}$$

This procedure is valid not only for luminous but also arbitrary radiant flux, so the expressions have general validity. Note that calculating actual view factors for real problems may be extremely complicated.

8.4 Light sources

8.4.1 Position of the Sun

The most important light source on the Earth is the Sun. The Sun provides us with the sense of time and seasons, as its apparent position in the sky defines basic time periods, year and day. However, to take the Sun into the account, we need two quantities that describe the position or object in the sky (Fig. 8.15):

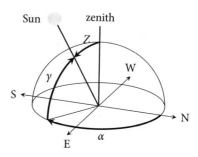

Figure 8.15: Azimuth α, angle of elevation γ and zenith angle Z for the position or object in the sky. For the Sun, all symbols bear index 's'.

- *Azimuth* α is the (clockwise) angle between the horizontal projection of the line passing through the object and the north direction ($\alpha = 0°$ is north, $\alpha = 90°$ is east, $\alpha = 180°$ is south, $\alpha = 270°$ is west).

- *Angle of elevation* or *altitude* γ is the angle between the line passing through the object and the horizontal plane.

- *Zenith angle* Z is the angle between the line passing through the object and zenith.

There are several quasi-empirical expressions for calculating the azimuth α_s and angle of elevation γ_s of the Sun as a function of the day of the year J ($J = 1$ is 1st January and $J = 365$ is 31st December) and astronomical time t in hours ($t = 12\,\mathrm{h}$ is noon). Here we will follow the procedure prescribed in EN 17037 [19]. The basis of the azimuth and angle of elevation calculation is *solar declination* δ of the Earth, that is, the angle between the rays of the Sun and the plane of the Earth's equator. Solar declination is a consequence of the Earth's axial tilt of 23.44°, that is, the angle between the Earth's axis and a line perpendicular to the Earth's orbit. Because the Earth revolves around the Sun, declination changes and can be described by the approximative expression

$$\delta = 0.3948° - 23.2559° \cos(J' + 9.1°) -$$
$$- 0.3915° \cos(2J' + 5.4°) - 0.1764° \cos(3J' + 26.0°), \qquad (8.34)$$

where $J' = 360° J/365$. The angle 9.1° corresponds to $J = -9.2$ and the winter solstice, that is the moment when the solar declination reaches one of the extreme values. Note that the declination value is the same for the whole Earth.

We also have to define *hour angle* ω_η, which describes the daily change of the Sun's apparent position. The hour angle is expressed as

$$\omega_\eta = \frac{360°}{24\,\mathrm{h}}(t - 12\,\mathrm{h}). \qquad (8.35)$$

The Sun is in the highest apparent daily position for $t = 12\,\mathrm{h}$, which means that the astronomical time depends on the geographic longitude of the observer. Note that the astronomical time is not necessarily equal to the local standard time. Angle of 1° between the meridian of the observer and the standard time meridian corresponds to a difference of 4 min.

The azimuth and angle of elevation can be calculated using the expressions

$$\gamma_s = \arcsin(\cos \omega_\eta \, \cos \varphi \, \cos \delta + \sin \varphi \, \sin \delta), \qquad (8.36)$$

$$\alpha_s = \begin{cases} 180° - \arccos \dfrac{\sin \gamma_s \, \sin \varphi - \sin \delta}{\cos \gamma_s \, \cos \varphi}, & t \leq 12 \, \text{h}, \\[2mm] 180° + \arccos \dfrac{\sin \gamma_s \, \sin \varphi - \sin \delta}{\cos \gamma_s \, \cos \varphi}, & t > 12 \, \text{h}, \end{cases} \qquad (8.37)$$

where φ is the geographical latitude.

In addition to the annual north-south oscillation of the Sun's apparent position due to declination, there is also a smaller but more complex annual east-west oscillation. This is the consequence of the elliptical shape of the Earth's orbit around the Sun. This essentially means that in terms of astronomical time (which assumes that all days are equally long), the Sun won't necessarily achieve its largest angle of elevation at $t = 12 \, \text{h}$. We correct that fact by an *equation of time*, a correction to astronomical time whose approximate expression can be written down as

$$\begin{aligned} \Delta t = {} & 0.000\,11 \, \text{h} + 0.122\,54 \, \text{h} \cdot \cos(J' + 85.9°) + \\ & + 0.165\,60 \, \text{h} \cdot \cos(2\,J' + 108.9°) + \\ & + 0.005\,65 \, \text{h} \cdot \cos(3\,J' + 105.2°). \end{aligned} \qquad (8.38)$$

The angle 85.9° corresponds to $J = -87.1$ and the half time between the Earth perihelion (position closest to the Sun) and the aphelion (position farthest from the Sun). The angle $(108.9° - 90°)/2$ corresponds to $J = -9.6$ and the winter solstice.

The preceding expressions are only approximative, not taking into account less important and more complex effects, such as atmospheric refraction and imperfectness of the Earth's shape compared to the ideal sphere.

Azimuth and angle of elevation are presented in the sun path diagram. They can be presented in

- the cylindrical sun path diagram, which uses the Cartesian coordinate system;

- the polar sun path diagram, which uses the polar coordinate system; and

- the stereographic sun path diagram, which uses the stereographic projection (circle radii are not equidistant) and corresponds to photographs taken with a fisheye lens.

Figure 8.16 presents the cylindrical sun path diagram for geographical latitude $\varphi = 47°$ (Northern Hemisphere). The top curve represents summer solstice, the bottom curve winter solstice and the middle curve equinox movement of the Sun. As explained previously, for a fixed astronomical time, the Sun's apparent position changes annually in terms of both azimuth and angle of elevation, creating typical figure eight loops called *analemmas*.

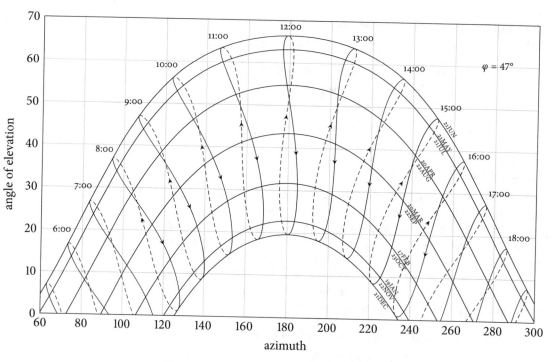

Figure 8.16: Sun path diagram. The top curve represents summer solstice, bottom curve winter solstice and middle curve equinox movement of the Sun. Fifteen loopy curves represent analemmas for astronomical times on the hour between 5:00 and 19:00 with dashed lines going up for the first half of the year and full lines going down for the second half.

Sun path diagrams in engineering are used to determine whether a position of interest (usually the façade of a building) has sunlight exposure at the specified time. First obstructions are identified and plotted on the diagram. To do that, azimuth and angle of elevation pairs representing the outline of the obstructions are determined and then plotted and connected on the diagram. The remaining area of the diagram represents sunlight exposure. Sun path diagrams can also be used to create precise sundials.

Example 8.2: Sunlight exposure for the building.

At distance l = 10 m south of the existing building, positioned on the horizontal surface at geographical latitude 47° (Northern Hemisphere), two new buildings in the shape of a cuboid of width d = 20 m and height h = 12.5 m are planned, as shown in the picture. Calculate the ground floor sunlight exposure for the existing building at solstices and equinox. Take the middle of the façade at height h_0 = 1.5 m above ground as the reference position.

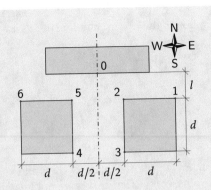

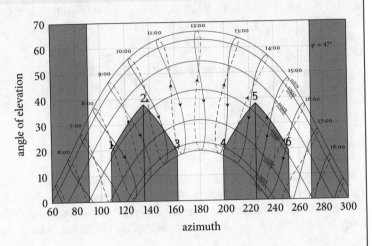

First, note that the façade of the existing building itself obstructs the view north of the east direction ($\alpha < 90°$) and north of the west direction ($\alpha > 270°$).

Next, we want to find the new buildings' obstruction region. Three extremal points suffice to outline each of the buildings, as shown in the picture. The horizontal distances from the reference point to the eastern building points are

$$d_1 = \sqrt{l^2 + \left(\tfrac{3}{2}d\right)^2} = 31.6\,\text{m},$$

$$d_2 = \sqrt{l^2 + \left(\tfrac{1}{2}d\right)^2} = 14.1\,\text{m},$$

$$d_3 = \sqrt{(l+d)^2 + \left(\tfrac{1}{2}d\right)^2} = 31.6\,\text{m}.$$

Azimuths of the eastern building points are (calculated relatively to the south direction, for which azimuth is 180°):

$$\alpha_1 = 180° - \arctan \frac{\tfrac{3}{2}d}{l} = 108°,$$

$$\alpha_2 = 180° - \arctan \frac{\tfrac{1}{2}d}{l} = 135°,$$

$$\alpha_3 = 180° - \arctan \frac{\tfrac{1}{2}d}{l+d} = 162°.$$

Angles of elevation of the eastern building points are

$$\gamma_1 = \arctan \frac{h - h_0}{d_1} = 19°,$$

$$\gamma_2 = \arctan \frac{h - h_0}{d_2} = 38°,$$

$$\gamma_3 = \arctan \frac{h - h_0}{d_3} = 19°.$$

We draw the points in the sun path diagram and connect them. Note that, in reality, connection curves are not lines, but for short distances in the sun path diagram, they can be approximated that way. The area below the curves presents the obstruction region of the eastern building. The same procedure can be repeated for the western building.

We conclude that sunlight exposure is about 2.5 h at the winter solstice, about 9 h at the summer solstice and about 7 h at the equinox.

8.4.2 Natural light sources

As already pointed out, the most important natural light source is the Sun. The Sun not only illuminates directly but also indirectly through clouds, atmosphere and reflections from earthly objects. The starting point of this study is therefore the sky as the combined effect of the Sun, clouds and atmosphere.

Because skylight source distance is nonuniform and difficult to determine, the specification of the luminous flux or luminous intensity is not very helpful. Instead, we can use luminance and calculate illuminance using expression (8.24).

Standard ISO 15469 [47] defines 16 sky models for various meteorological conditions (for example, clear sky, cloudy sky and overcast sky) based on studies by the International Commission on Illumination (Commission Internationale de l'Eclairage, CIE). For most sky models, luminance expressions are complex and dependent on the Sun's apparent position. Some simpler models are also defined, such as the *CIE standard overcast sky* and *CIE traditional overcast sky*, where luminance depends only on the angle of elevation. The expression for the latter is

$$L_v(\gamma) = L_Z \frac{1 + 2\sin\gamma}{3}, \tag{8.39}$$

where L_Z is the luminance at the zenith. An even simpler model is the uniform sky with

$$L_v(\gamma) = L_Z. \tag{8.40}$$

Absolute values for the luminance at the zenith generally depend on the solar elevation. The standard does not prescribe any expression, but empirical expressions have been published elsewhere for different climates.

Here we present one useful formula for demonstration purposes:

$$L_Z = \frac{9}{7\pi}(300 + 21000 \sin \gamma_s),\qquad (8.41)$$

where γ_s is the angle of elevation of the Sun.

Example 8.3: Illuminance under standard sky.

Calculate the illuminance of a small horizontal surface under a CIE traditional overcast sky and uniform sky.

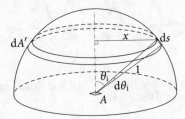

Because luminance depends on the direction, the calculation is done using expression (8.24) in the differential form of

$$E_v = \int L_v \cos \theta_i \, d\Omega.$$

For the CIE standard overcast sky, in which luminance depends only on the angle of incidence θ_i, the half-space has to be split into small differential solid angles of constant θ_i, as shown in the figure. Because the radius is equal to $r = 1$, we get

$$d\Omega = 2\pi x \, ds = 2\pi \sin \theta_i \, d\theta_i.$$

Putting this and (8.39) into the first expression, we get

$$E_v = 2\pi L_Z \int_0^{\pi/2} \frac{1 + 2\cos \theta_i}{3} \cos \theta_i \sin \theta_i \, d\theta_i = \frac{7\pi}{9} L_Z.$$

Combining the result with (8.41) gives

$$E_v = 300 + 21000 \sin \gamma_s,$$

which is the horizontal illuminance from the unobstructed sky, proposed by Krochmann et al. [73].

On the other hand, for the uniform sky, we get

$$E_v = 2\pi L_Z \int_0^{\pi/2} \cos \theta_i \sin \theta_i \, d\theta_i = \pi L_Z.$$

Even objects without sky access are illuminated during the day. This is due to reflected skylight from earthly objects, which are also considered a natural light source.

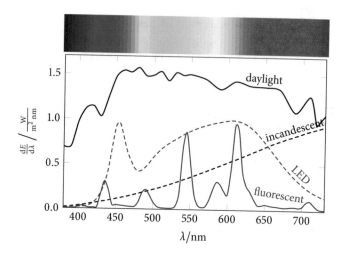

Figure 8.17: Daylight spectral irradiance [33] and typical spectral irradiances of three artificial sources [81].

8.4.3 Artificial light sources

When natural light sources do not satisfy our needs for illumination, artificial sources should be used.

In early history, humans used fire as a means of illumination, whereas more sophisticated sources such as candles and oil lamps were invented in antiquity. The Industrial Age brought illumination by gas lamps and finally incandescent light bulbs. The common property of all these light sources is that they mimic the light production by the Sun. The source of light is black body radiation due to high temperatures. In candles, oil lamps and gas lamps, air is heated by burning a particular fuel. In incandescent light, the bulb filament wire is heated by electric current (Fig. 2.20 on page 55).

In all described cases, the temperatures are considerably smaller than the temperature of the Sun, which means that the spectral density of the heat flow rate function (Fig. 2.19 on page 54) peaks in the infrared region. This has two important consequences:

1. These light sources are much less efficient than the Sun, because the proportion of visible light in the total radiation is much larger than the proportion of infrared radiation. Whereas the luminous efficacy of the Sun is $K = 93 \, \text{lm/W}$, the luminous efficacy of an incandescent light bulb is only about $K = 20 \, \text{lm/W}$.

2. These light sources emit a larger share of red light and a smaller share of blue light compared to the Sun, which means that the emitted colour of the mixture is different (Fig. 8.17). Because these sources are approximately black body radiators, we can characterise the emitted colours by their temperature, which is called *colour temperature* T_c for the purpose.

In recent decades, new types of artificial light sources, such as fluorescent lamps and light emitting diode (LED) are slowly displacing older

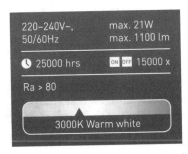

220–240V~,   max. 21W
50/60Hz   max. 1100 lm

🕒 25000 hrs   ON OFF 15000 x

Ra > 80

3000K Warm white

Figure 8.18: Specifications on the package of the light product. Along with important light-related specifications, such as luminous flux, correlated colour temperature and colour rendering index, the electrical voltage and power, and product durability are specified.

types of light sources. The light is produced at room temperatures when electrons jump between two quantum energetic states. Because these types of sources produce the radiation almost exclusively in the visual spectrum, they are much more efficient, with luminous efficacy of about $K = 70\,\text{lm/W}$ for fluorescent lamps and $K = 90\,\text{lm/W}$ for LEDs. The *correlated colour temperature* T_{cp} of these light sources is obtained by comparing the emitted light spectrum with the spectrum of black bodies at various temperatures until the best match is obtained. Colour temperature and correlated colour temperature are standard specifications of light products (Fig. 8.18).

As pointed out before, light sources at higher (correlated) colour temperatures produce a larger share of blue light and a smaller share of red light, whereas sources at lower (correlated) colour temperatures produce a larger share of red light and a smaller share of blue light. This leads to a paradox, because it is opposite to the cultural associations attributed to colours, in which red is 'warm', and blue is 'cold': Light sources at higher colour temperatures produce 'cold' light, and light sources at lower colour temperatures produce 'warm' light. Colour appearance groups are listed in Table 8.1.

The old types of light sources emit a continuous colour spectrum, whereas fluorescent lamps and LEDs emit separate spectral lines, that is, separate colours (Fig. 8.17). Those colours are well mixed together to give the appearance of the white light. But new types of artificial light sources perform poorly when their light is reflected by objects, whose colour we want to inspect. To quantify colour appearance of objects under different light sources, CIE *general colour rendering index* (CRI) R_{a} is defined. The

Table 8.1: Lamp colour appearance groups [13]. Cultural conventions lead to a paradox that sources with lower (correlated) temperature produce a warmer light.

Colour Appearance	T_{cp}/K
warm	<3300
intermediate	3300–5300
cool	>5300

highest possible value is 100, which corresponds to a light source that is identical to standardised daylight. As the colour appearance drifts away from the daylight ideal, the value gets smaller, dropping to negative values for some light sources. Despite the fact that the colour temperature of new types of light sources have larger correlated colour temperatures that are closer to the colour temperature of daylight, they usually have a lower colour rendering index. CRI is one of the standard specifications of the light products (Fig. 8.18).

8.5 Calculation of illuminance

The main concern regarding building is illuminance of objects both inside and outside of buildings. In calculations, we have to take into account not only light coming from natural and artificial light sources, but also reflections from surfaces, which are treated as additional light sources. In order to calculate total indoor illuminance E_i using the *split-flux method*, contributions from three types of sources are added together (Fig. 8.19):

1. *Sky component* (SC) is the direct light from a patch of sky visible at the point considered.

2. *Externally reflected component* (ERC) is the light reflected from an external surface and then reaching the point considered.

3. *Internally reflected component* (IRC) is the light entering through the window but reaching the point considered only after reflection from an internal surface.

Other calculation methods include the following [74]:

- *Radiosity* is an algorithm in which surfaces are split into smaller patches using a finite element method, and a view factor (Section 8.3.4) is calculated for each pair of patches. Illumination is then calculated by adding the contributions from all visible surrounding patches and light sources.

- *Ray tracing* is a technique that calculates the distribution of a large number of rays emitted either from light sources (forward ray tracing) or of rays striking a point of interest (backward ray tracing).

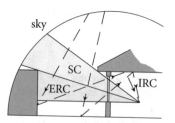

Figure 8.19: Illuminance components according to the split-flux method: sky component (SC), externally reflected component (ERC) and internally reflected component (IRC).

Table 8.2: Ranges of useful reflectances for the major interior diffusely reflecting surfaces [13, 19].

Surface	Reflectance
ceiling	0.7–0.9
interior walls	0.5–0.8
floor	0.2–0.4
exterior walls	0.2–0.4
exterior ground	0.2

CIE standard overcast sky at a particular day is usually relevant for these calculations.

Furthermore, standard EN 12464-1 [13] defines ranges of useful reflectances for the major internal surfaces (Table 8.2).

Note that calculating illuminance is a very tedious and complex task and is therefore usually done by special computer programs.

In order to evaluate illumination quality, additional illuminance-related quantities are considered and calculated:

- *Illuminance uniformity* U_o is the ratio of minimum illuminance E_{min} to average illuminance E_{av} on a surface [15]:

$$U_o = \frac{E_{min}}{E_{av}}.$$

- *Daylight factor D* is the ratio of indoor illuminance E_i to the outdoor illuminance E_o, excluding the direct sunlight contribution:

$$D = \frac{E_i}{E_o}. \tag{8.42}$$

It is normally determined by numerical calculation of the illuminances using CIE standard overcast sky or CIE traditional overcast sky (Section 8.4.2) for the reference plane 0.85 m above the floor (Fig. 8.20) [19]. The result does not depend on the selected calculation time, but usually the Solar noon on the equinox is used. Sometimes also approximate equations, protractors, nomograms, diagrams and tables are used.

A special concern should be *glare*, which is defined as meeting at least one of two conditions:

- too much light; or

- excessive contrast, meaning the range of luminance in the field of view is too great.

The CIE *unified glare rating* R_{UG} is defined to evaluate glare for indoor working places [13]. It is defined as

$$R_{UG} = 8 \lg \left[\frac{0.25}{L_b} \sum_n \left(L_n^2 \frac{\Omega_n}{p_n^2} \right) \right],$$

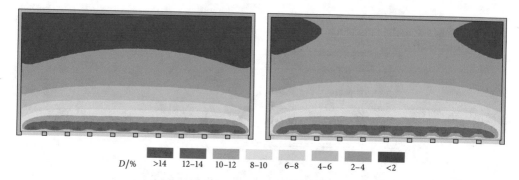

$D/\%$ >14   12–14   10–12   8–10   6–8   4–6   2–4   <2

Figure 8.20: Daylight factor for a lecture room with 10 windows (left) and the same lecture room with a light shelf, as shown in Fig. 8.21 (right). In the presented case, a light shelf extends the daylighted zone depth from approximately 1.5 to 2.5 times the window head height. In addition, the illuminance uniformity of the room is increased [75].

where L_b is the background luminance, and L_n, Ω_n and p_n are the luminance, solid angle and Guth position index of each light source, respectively. The Guth position index is a number that increases with distance from direct line of sight. It is based on two angles: α is the angle from the vertical of the plane containing the source and the line of sight in degrees, and β is the angle between the line of sight and the line from the observer to the source. The Guth position index is expressed as

$$p = \exp\left[\left(35.2 - 0.318\,89\,\alpha - 1.22\exp\left(-\tfrac{2}{9}\alpha\right)\right)10^{-3}\beta + \right.$$
$$\left. + \left(21 + 0.266\,67\,\alpha - 0.002\,936\,\alpha^2\right)10^{-5}\beta^2\right].$$

On the other hand, CIE *glare rating* R_G is defined to evaluate glare for outdoor working places [14].

8.6 Requirements regarding building illumination

Illumination must fulfil several conditions:

1. Buildings must have appropriate sunlight exposure. The requirements are usually expressed in number of hours of direct sunlight per day for a clear cloudless day, usually near the winter solstice. The standard EN 17037 [19] stipulates that the minimum sunlight exposure for patient rooms in hospitals, play rooms in nurseries and at least one habitable space in the dwelling on a selected date between 1st February and 21st March is 1.5 h. The reference point is located on the inner surface of the opening, horizontally centred, at least 1.2 m above the floor and 0.3 m above the sill of the opening. Additionally, sunlight exposure below the minimum solar angle of elevation (specified for European countries) should be rejected. Note that the calculation examples in this book, which use a reference point on the façade, are therefore simplified.

2. Working surfaces and other areas must have appropriate illuminance, depending on the activity. Standard EN 12464-1 [13] defines
maintained illuminance E_m, which is the value below which the average illuminance is not allowed to fall. For most indoor working
places, illuminance should be between 200 lx and 500 lx. For certain
types of rooms (for example, corridors, entrance halls, restrooms),
illumination of 100 lx suffice. For light-sensitive work (for example,
medical examination and operation, art rooms in art schools, technical drawing, precision work, colour and other inspection), illuminance should be between 750 lx and 1500 lx. In some very special
cases, up to 5000 lx might be required. On the other hand, according
to standard EN 12464-2 [14], illumination of 10 lx to 200 lx suffices
in most outdoor working places.

3. A building must have sufficient daylight throughout the year. The
standard EN 17037 [19] provides for two methods of daylight evalu-
ation:

- The calculation of the illumination for the reference plane
  0.85 m above the floor. For daylight openings in vertical and
  inclined surfaces the illuminance must exceed two values: the
  target illuminance E_T = 300 lx for at least 50 % of space and
  the minimum target illuminance E_{TM} = 100 lx for at least 95 %
  of space. For daylight openings in horizontal surfaces, the illuminance must exceed the target illuminance E_T = 300 lx for
  at least 95 % of space. The calculation must be carried out for
  4380 hours of daylight throughout the year and the above conditions must be met for at least 50 % of hours.

- The calculation of a single daylight factor (8.42) can replace the
  calculation of multiple hour internal illuminances, provided
  that we know the minimum value of the largest 50 % of the
  hour external illuminances: the median external diffusive illuminance $E_{v,d,med}$, which excludes the direct Sun contribution,
  or the median external global illuminance $E_{v,g,med}$, which includes the direct Sun contribution. For daylight openings in
  vertical and inclined surfaces, the minimum daylight factors
  are

$$D_T = \frac{E_T}{E_{v,d,med}}, \quad D_{TM} = \frac{E_{TM}}{E_{v,d,med}}.$$

  Note that smaller median external diffusive illuminance, usually at larger geographical latitudes, requires a larger daylight
  factor to produce the same daylight illuminance. The standard also lists the median external diffusive illuminances and
  the minimum daylight factors for European countries. In particular, the median external diffusive illuminance in Europe
  is $E_{v,d,med}$ = 11 500 lx–19 400 lx, so that the daylight factor
  must exceed D_T = 1.5 %–2.6 % for at least 50 % of space and
  D_{TM} = 0.5 %–0.9 % for at least 95 % of space. For daylight
  openings in horizontal surfaces, the Sun direct contribution
  is taken into account, so that instead of the median diffusive

global illuminance, the median global diffusive illuminance is used.

The standard EN 17037 [19] also gives recommendations for the view out and the glare protection, which is beyond the scope of this book.

4. Illuminance uniformity should be appropriate. Standard EN 12464-1 [13] prescribes that minimum illuminance uniformity U_o for indoor working places should be in most cases between 0.4 and 0.7. On the other hand, standard EN 12464-2 [14] prescribes that minimum U_o for the outdoor working places should be in most cases between 0.25 and 0.5.

5. Glare should be limited within acceptable levels. Standard EN 12464-1 [13] prescribes a maximum value of unified glare rating limit R_{UGL}. Its value is 19 for most indoor working places, whereas for glare-sensitive work (for example, technical drawing, precision work, control rooms, colour inspection), its value is 16.

6. The colour rendering index should be as large as possible. Standard EN 12464-1 [13] prescribes that colour rendering index R_a should be at least 80 for most indoor working places, whereas for colour-sensitive work (for example, most healthcare premises, art rooms in art schools, colour inspection), it should be at least 90. On the other hand, standard EN 12464-2 [14] prescribes that minimum R_a for outdoor working places should be in most cases between 20 and 40.

The daylight factor can be improved by

- increasing externally reflected components by using light colours for façades of surrounding buildings and

- increasing the sky component by adding daylight openings and placing them in a way that they are oriented more towards the unobstructed overcast sky.

Uniformity of illumination and depth of daylight penetration can be improved by

- increasing window head height h (Fig. 8.21),

- using light shelves with high ceilings (Fig. 8.21),

- increasing internally reflected components by using light colours for room surfaces, especially in the rear half of the room,

- using highly reflective windows sills,

- placing windows near reflective surfaces (walls), and

- using skylights or light tubes to bring light in from above.

Generally, light shelves can expand into the exterior, interior or both. They are often optimised not to increase the daylighted zone (calculated for the CIE standard overcast sky) but to illuminate deeper parts of the room with the sunlight. In this case, the light shelf design must also take into account the sun path.

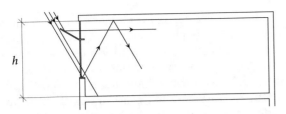

Figure 8.21: A light shelf and a reflective window sill transmits light deeper into the room. The effect on the daylight factor is shown in Fig. 8.20. The daylighted zone depends primarily on window head height h.

Glare control depends on the source of the glare:

- Glare due to direct natural light is reduced by the use of window shades and ground glass.

- Glare due to direct artificial light is reduced by using light fixtures, which either direct light away from the working area (usually towards the ceiling) or obstruct the light path with ground glass.

- Glare due to reflections is reduced by using diffusive surfaces.

Problems

8.1 Four identical vertical street lamps of height $h = 8.0$ m are positioned in a way that their projection on the ground forms a square with sides of $a = 5.0$ m, as shown in the figure. Calculate the luminous intensity of each isotropic light so that illuminance in the centre of the square (point 0) is 50 lx. (1050 cd)

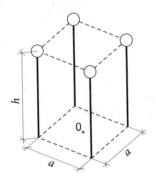

8.2 Calculate the solid angle of the Sun observed from the Earth, if the Sun is about 1.50×10^8 km away from the Earth, and the Sun's radius is 6.96×10^5 km. Assuming that the luminance of the Sun on the Earth's surface is 1.44×10^9 cd/m^2 at zenith on a sunny day, calculate the maximum illuminance of the small horizontal surface. (6.76×10^{-5} sr, 9.74×10^4 lx)

8.3 Two cuboid-shaped buildings of height 9.5 m are positioned on the horizontal surface at geographical latitude 47° (Northern Hemisphere), as shown in the figure. Buildings are rotated with respect to east-west direction for $\theta = 30°$. Using a sun path diagram, determine the sunlight exposure on 21st March for a window at two-thirds of the south façade (point 0) at height 1.5 m. For the calculation, consider only the angles of elevation of the Sun that are larger than 15° and use the equidistant points 1 to 4. (3 h 20 min)

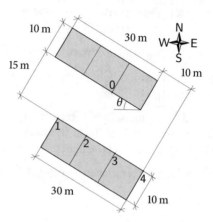

8.4 Isotropic light of luminous flux 1500 lm is located 2.2 m above the floor and 2.0 m from the vertical wall with the mirror, as shown in the figure. Calculate the illumination of the floor 1.0 m from the wall. Take into account only the direct light and the light reflected by the mirror. (23.7 lx)

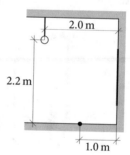

A Tables

Symbols and names of the building physics quantities comply with standard ISO 80000, ISO 7345 and EN 12665 [53, 54, 55, 56, 57, 58, 59, 60, 28, 15]. Sometimes, more common expressions are used, marked with †.

Table A.1: Symbols and names of building physics quantities.

Symbol	Unit	Name
A	J	work
A	m^2	area
A	m^2	equivalent absorption area
A	dB	sound attenuation
a	m/s^2	acceleration
a	m^2/s	thermal diffusivity
b	$W\,s^{\frac{1}{2}}/(m^2\,K)$	thermal effusivity
c	m/s	(phase) speed
c	$J/(kg\,K)$	specific heat capacity
c_0	m/s	speed of light in vacuum
c_p	$J/(kg\,K)$	specific heat capacity at constant pressure
c_V	$J/(kg\,K)$	specific heat capacity at constant volume
D	m^2/s	water vapour diffusion coefficient
D	dB	level difference
D	1	daylight factor
D_C	dB	directivity correction
E	J	mechanical energy
E	W/m^2	irradiance
E_v	lx	illuminance
e	J/m^3	time-averaged density of mechanical energy
F	1	view factor
F	N	force
f	Hz	frequency
f	1	fractional area
$f_{R_{si}}$	1	temperature factor of the internal surface
g	m/s^2	acceleration of free fall
g	$kg/(m^2\,s)$	density of water vapour flow rate
g	1	solar factor
H	J	enthalpy
H	W/K	heat transfer coefficient
h	J s	Planck's constant
h	J/kg	specific enthalpy

Table A.1: Symbols and names of building physical quantities (continued)

Symbol	Unit	Name
h	W/(m² K)	surface coefficient of heat transfer
h_c	W/(m² K)	convective surface coefficient
h_f	J/kg	specific enthalpy of fusion
h_m	m/s	surface coefficient of mass transfer
h_r	W/(m² K)	radiative surface coefficient
h_v	J/kg	specific enthalpy of vaporisation
I	W/m²	time-averaged sound intensity
I_v	cd	luminous intensity
i	W/m²	sound intensity
K	N/m²	modulus of compression
K	s	liquid water conductivity
K	dB	level adjustment
K_m	lm/W	maximum spectral luminous efficacy
k	rad/m	angular wavenumber
k	W/(m² K)	coefficient of heat transfer
k	J/K	Boltzmann's constant
L_i	dB	impact sound pressure level
L_p	dB	sound pressure level
L_R	dB	rating level
L_v	cd/m²	luminance
L_W	dB	sound power level
L'_W	dB	linear sound power level
L''_W	dB	surface sound power level
l	m	length
M	kg/mol	molar mass
M_v	lm/m²	luminous exitance
m	kg	mass
m	kg	mass of water vapour
m_d	kg	mass of dry air
N_A	1/mol	Avogadro's constant
n	mol	amount of substance
n	1/s	air change rate
n_{pr}	1/s	air change rate at the reference pressure difference
P	W	power
P	W	sound power
P'	W/m	linear sound power
P''	W/m²	surface sound power
p	Pa	pressure
p	Pa	pressure of water vapour
p	Pa	sound pressure
p_{atm}	Pa	atmospheric pressure
p_d	Pa	pressure of dry air
p_s	Pa	static pressure
p_{sat}	Pa	water vapour pressure at saturation
Q	J	heat
q	W/m²	density of heat flow rate
q_f	J/kg	specific heat of fusion

Table A.1: Symbols and names of building physical quantities (continued)

Symbol	Unit	Name
q_m	kg/s	mass flow rate
q_{pr}	m^3/s	air leakage rate at the reference pressure difference
q_V	m^3/s	air flow rate
q_v	J/kg	specific heat of vaporisation
R	dB	sound reduction index
R	m^2 K/W	thermal resistance
R	J/(mol K)	molar gas constant
R_a	1	colour rendering index
R_a	m^2 K/W	thermal resistance of airspace
R_{se}	m^2 K/W	external surface resistance
R_{si}	m^2 K/W	internal surface resistance
R_{tot}	m^2 K/W	total thermal resistance
R_{UG}	1	unified glare rating
R_v	J/(kg K)	gas constant for water vapour
S	m^2	area
S	m/s^{½}	sorptivity
S	W/m^2	Poynting's vector
s_d	m	water vapour diffusion - equivalent air layer thickness
T	K	temperature
T	s	period
T_c	K	colour temperature
T_{cp}	K	correlated colour temperature
T_n	s	reverberation time
t	s	time
U	J	internal energy
U	W/(m^2 K)	thermal transmittance
U_o	1	uniformity
u	m^3	mass ratio of water to dry matter *water (moisture) content* †
V	m^3	volume
V	1	spectral luminous efficiency
v	m/s	velocity
v	kg/m^3	mass concentration of water vapour *absolute humidity* †
w	kg/m^3	mass concentration of water
x	m	displacement
x	1	mass ratio of water vapour to dry gas *humidity ratio* †
Z	1	zenith angle
α	1	absorptance
α	1	absorbance
α	1	azimuth
α_l	1/K	linear expansion coefficient
α_V	1/K	cubic expansion coefficient
γ	N/m	surface tension
γ	1	ratio of heat capacities

Table A.1: Symbols and names of building physical quantities (continued)

Symbol	Unit	Name
γ	1	angle of elevation or altitude
δ	m	periodic penetration depth
δ	1	dissipance
δ_0	$kg/(m\,s\,Pa)$	water vapour permeability with respect to vapour pressure
ε	1	emittance
η	1	efficiency
θ	°C	temperature
θ	1	contact angle
λ	m	wavelength
λ	$W/(m\,K)$	thermal conductivity
μ	1	water vapour resistance factor
ρ	kg/m^3	density
ρ	1	reflectance
ρ_A	kg/m^2	surface density
ρ_l	kg/m	linear density
σ	$W/(m^2\,K^4)$	Stefan-Boltzmann constant
τ	1	transmittance
Φ	W	heat flow rate
Φ	W	radiant flux
Φ_v	lm	luminous flux
φ	rad	phase constant
φ	1	relative humidity
χ	W/K	point thermal transmittance
Ψ	$W/(m\,K)$	linear thermal transmittance
ψ	1	moisture content volume by volume
Ω	sr	solid angle
ω	$1/s$	angular frequency

Notes: Adjective *specific* is added to the quantity name to designate the quotient of that quantity by *mass*. Units for angles (zenith angle, azimuth, angle of elevation and contact angle) are usually expressed in degrees, but are essentially dimensionless.

Table A.2: Physical constants.

Quantity	Symbol	Value
speed of sound in air at 20 °C	c	343.2 m/s
speed of light in vacuum	c_0	2.998×10^8 m/s
standard acceleration of free fall	g	9.807 m/s^2
Planck's constant	h	6.626×10^{-34} J s
maximum spectral luminous efficacy	K_m	683 lm/W
Boltzmann's constant	k	1.381×10^{-23} J/K
sound power reference value	P_0	1.0×10^{-12} W
standard atmospheric pressure	p_{atm}	1.013×10^5 Pa
sound pressure reference value	p_0	2.0×10^{-5} Pa
molar gas constant	R	8.314 J/(mol K)
Avogadro's constant	N_A	6.022×10^{23}/mol
gas constant for water vapour	R_v	461.5 J/(kg K)
water vapour permeability	δ_0	2×10^{-10} kg/(m s Pa)
with respect to vapour pressure of air		
standard density of liquid water	ρ_0	1.00×10^3 kg/m^3
Stefan Boltzmann constant	σ	5.670×10^{-8} W/(m^2 K^4)

Table A.3: Physical properties of the most common building-related materials. Presented values are typical or averaged values from ISO 10456 [40].

Material	$\rho / \frac{kg}{m^3}$	$c / \frac{J}{kg\,K}$	$c\rho / \frac{kJ}{m^3\,K}$	$\lambda / \frac{W}{m\,K}$	μ
expanded polystyrene	30	1450	43.5	0.035	60
mineral wool	100	1030	103	0.035	1
timber	500	1600	800	0.13	50
brick, solid †	1800	1000	1800	0.8	16
brick, perforated †	680	1000	680	0.22	8
gypsum plasterboard	700	1000	700	0.21	10
concrete, medium density	2200	1000	2200	1.65	120
soda lime glass	2500	750	1880	1.0	∞
ceramic, porcelain	2300	840	1930	1.3	∞
steel	7800	450	3510	50	∞
polyethylene	950	2000	1900	0.42	100 000
stone, crystalline	2800	1000	2800	3.5	10 000
stone, sedimentary	2600	1000	2600	2.3	250
soil, sand/gravel	1950	1050	2050	2.0	50
soil, clay/silt	1500	2090	3140	1.5	50
ice	900	2000	1800	2.2	–
liquid water, still	1000	4190	4190	0.6	–
air, still	1.23	1008‡	1.24	0.025	1

† values obtained from producers

‡ at constant pressure

Bibliography

[1] T.L. Bergman, A.S. Lavine, F.P. Incropera, D.P. Dewitt. *Fundamentals of Heat and Mass Transfer, 7th edition*. John Wiley & Sons, 2011.

[2] R.A. Serway in J.W. Jewett. *Physics for Scientists and Engineers, 8th edition*. Brooks/Cole, Cengage Learning, Belmont, 2009.

[3] S. Medved, *Gradbena fizika*. Fakulteta za arhitekturo, Ljubljana, 2010.

[4] V. Šimetin, *Građevinska fizika*. Fakultet građevinskih znanosti, Zagreb, 1983.

[5] ANSI/ASHRAE Standard 62.2-2016, Ventilation and Acceptable Indoor Air Quality in Residential Buildings.

[6] ASTM E779 – 03, Standard Test Method for Determining Air Leakage Rate by Fan Pressurization.

[7] DIN 4108-7:2011-01, Wärmeschutz und Energie-Einsparung in Gebäuden – Teil 7: Luftdichtheit von Gebäuden – Anforderungen, Planungs- und Ausführungsempfehlungen sowie beispiele.

[8] DIN 4109-1:2016-07, Schallschutz im Hochbau — Teil 1: Mindestanforderungen.

[9] DIN 4109-2:2016-07, Schallschutz im Hochbau — Teil 2: Rechnerische Nachweise der Erfüllung der Anforderungen.

[10] EN 410:2011, Glass in building – Determination of luminous and solar characteristics of glazing.

[11] EN 673:2011, Glass in building – Determination of thermal transmittance (U value) – Calculation method.

[12] EN 1991-1-5:2003, Actions on structures - Part 1-5: General actions - Thermal actions.

[13] EN 12464-1:2011, Lighting of work places – Part 1: Indoor work.

[14] EN 12464-2:2014, Lighting of work places – Part 2: Outdoor work places.

[15] EN 12665:2011, Light and lighting – Basic terms and criteria for specifying lighting requirements.

[16] EN 12898:2001, Glass in building – Determination of the emissivity.

© The Editor(s) (if applicable) and The Author(s), under exclusive license to Springer Nature Switzerland AG 2021
M. Pinterić, *Building Physics*,
https://doi.org/10.1007/978-3-030-67372-7

[17] EN 15026:2007, Hygrothermal performance of building components and building elements – Assessment of moisture transfer by numerical simulation.

[18] EN 16798-7:2017, Energy performance of buildings. Ventilation for buildings. Calculation methods for the determination of air flow rates in buildings including infiltration.

[19] EN 17037:2018, Daylight in buildings.

[20] IEC 61672-1:2013, Electroacoustics – Sound level meters – Part 1: Specifications.

[21] ISO 226:2003, Acoustics – Normal equal-loudness-level contours.

[22] ISO 354:2003, Acoustics – Measurement of sound absorption in a reverberation room.

[23] ISO 717-1:2013, Acoustics – Rating of sound insulation in buildings and of building elements – Part 1: Airborne sound insulation.

[24] ISO 717-2:2013, Acoustics – Rating of sound insulation in buildings and of building elements – Part 2: Impact sound insulation.

[25] ISO 1996-1:2016 Acoustics – Description, measurement and assessment of environmental noise – Part 1: Basic quantities and assessment procedures.

[26] ISO 1996-2:2017 Acoustics – Description, measurement and assessment of environmental noise – Part 2: Determination of environmental noise levels.

[27] ISO 6946:2007, Building components and building elements – Thermal resistance and thermal transmittance – Calculation method.

[28] ISO 7345:1987, Thermal insulation – Physical quantities and definitions.

[29] ISO 7730:2005, Ergonomics of the thermal environment – Analytical determination and interpretation of thermal comfort using calculation of the PMV and PPD indices and local thermal comfort criteria.

[30] ISO 9050:2003, Glass in building – Determination of light transmittance, solar direct transmittance, total solar energy transmittance, ultraviolet transmittance and related glazing factors

[31] ISO 9613-1:1993 Acoustics – Attenuation of sound during propagation outdoors – Part 1: Calculation of the absorption of sound by the atmosphere.

[32] ISO 9613-2:1996 Acoustics – Attenuation of sound during propagation outdoors – Part 2: General method of calculation.

[33] ISO 9845-1:1992, Solar energy – Reference solar spectral irradiance at the ground at different receiving conditions – Part 1: Direct normal and hemispherical solar irradiance for air mass 1.5.

[34] ISO 9972:2015, Thermal performance of buildings – Determination of air permeability of buildings – Fan pressurization method.

[35] ISO 10077-1:2017, Thermal performance of windows, doors and shutters – Calculation of thermal transmittance, Part 1: General.

[36] ISO 10140-2:2010, Acoustics – Laboratory measurement of sound insulation of building elements – Part 2: Measurement of airborne sound insulation.

[37] ISO 10140-3:2010, Acoustics – Laboratory measurement of sound insulation of building elements – Part 3: Measurement of impact sound insulation.

[38] ISO 10140-5:2010, Acoustics – Laboratory measurement of sound insulation of building elements – Part 5: Requirements for test facilities and equipment

[39] ISO 10211:2017, Thermal bridges in building construction – Heat flows and surface temperatures – Detailed calculations.

[40] ISO 10456:2007, Building materials and products – Hygrothermal properties – Tabulated design values and procedures for determining declared and design thermal values.

[41] ISO 12572:2001, Hygrothermal performance of building materials and products – Determination of water vapour transmission properties

[42] ISO 13370:2017, Thermal performance of buildings – Heat transfer via the ground – Calculation methods.

[43] ISO 13786:2017, Thermal performance of building components – Dynamic thermal characteristics – Calculation methods.

[44] ISO 13788:2012, Hygrothermal performance of building components and building elements – Internal surface temperature to avoid critical surface humidity and interstitial condensation – Calculation methods.

[45] ISO 13789:2017, Thermal performance of buildings – Transmission and ventilation heat transfer coefficients – Calculation method.

[46] ISO 14683:2017, Thermal bridges in building construction – Linear thermal transmittance – Simplified methods and default values.

[47] ISO 15469:2004, Spatial distribution of daylight – CIE standard general sky.

[48] ISO 16283-1:2014, Acoustics – Field measurement of sound insulation in buildings and of building elements – Part 1: Airborne sound insulation.

[49] ISO 16283-2:2015, Acoustics – Field measurement of sound insulation in buildings and of building elements – Part 2: Impact sound insulation.

[50] ISO 20065:2016, Acoustics – Objective method for assessing the audibility of tones in noise – Engineering method

[51] ISO 52019-2:2017, Energy performance of buildings – Hygrothermal performance of building components and building elements – Part 2: Explanation and justification.

[52] ISO 52022-2:2017, Energy performance of buildings – Thermal, solar and daylight properties of building components and elements – Part 2: Explanation and justification.

[53] ISO 80000-1:2013, Quantities and units, Part 1: General.

[54] ISO 80000-3:2012, Quantities and units, Part 3: Space and time.

[55] ISO 80000-4:2012, Quantities and units, Part 4: Mechanics.

[56] ISO 80000-5:2012, Quantities and units, Part 5: Thermodynamics.

[57] ISO 80000-7:2013, Quantities and units, Part 7: Light.

[58] ISO 80000-8:2013, Quantities and units, Part 8: Acoustics.

[59] ISO 80000-9:2013, Quantities and units, Part 9: Physical chemistry and molecular physics.

[60] ISO 80000-10:2013, Quantities and units, Part 10: Atomic and nuclear physics.

[61] Comission directive (EU) 2015/996 establishing common noise assessment methods according to Directive 2002/49/EC of the European Parliament and of the Council, 2015 O.J. L 168/1.

[62] M. Rubin, Sol. Energ. Mater. **12** (4), 275–288 (1985).

[63] Vitavia Europe ApS, Odense, Denmark.

[64] F. Hecht, J. Numer. Math. **20** (3-4), 251—265 (2012).

[65] Schöck Bauteile Ges.m.b.H., Wien, Austria.

[66] M.H. Sherman, Energy Build. **10**, 81—86 (1987).

[67] B. Djolani, Ann. Sci. Forest. **29** (4), 465–474 (1972).

[68] F. Pels Leusden, H. Freymark, Gesund. Ing. **72** (16), 271–273 (1951).

[69] W.D. Keidel, W.D. Neff, eds. *Handbook of Sensory Physiology (Vol. 1)*. Springer-Verlag, 1974.

[70] Epi Spektrum doo, Maribor, Slovenia.

[71] S. Jennings, J. Aerosol Sci. **19** (2), 159—166 (1988).

[72] G. Wyszecki, W.R. Blevin, K.G. Kessler, K.D. Mielenz, *Principles governing photometry*, BIPM, Monographie 83/1 (1983).

[73] J. Krochmann and M. Seidl, Light. Res. Technol. **6** (3) 165–171 (1974).

[74] A. Iversen, N. Roy, M. Hvass, M. Jørgensen, J. Christoffersen, W. Osterhaus, K. Johnsen. *Daylight calculations in practice*. Danish Building Research Institute, Aalborg University, 2013.

[75] Velux Daylight Visualizer.

[76] https://passipedia.org/basics/building_physics_-_basics/heating_load

[77] http://www.passiv.de/en/02_informations/02_passive-house-requirements/02_passive-house-requirements.htm

[78] https://buildingscience.com/documents/digests/bsd-138-moisture-and-materials

[79] http://www.anaesthesiamcq.com/FluidBook/fl3_2.php

[80] http://ec.europa.eu/environment/noise/europe_en.htm

[81] http://www.designingwithleds.com/light-spectrum-charts-data/

Index

Printed in the United States
by Baker & Taylor Publisher Services